ANIMAL AND VEGETABLE PROTEINS IN LIPID METABOLISM AND ATHEROSCLEROSIS

CURRENT TOPICS IN NUTRITION AND DISEASE

Series Editors

Anthony A. Albanese
The Burke Rehabilitation Center
White Plains, New York

David Kritchevsky
The Wistar Institute
Philadelphia, Pennsylvania

ANIMAL AND VEGETABLE PROTEINS IN LIPID METABOLISM AND ATHEROSCLEROSIS

Editors

Michael J. Gibney
Department of Nutrition
Medical School
University of Southampton
Southampton, England, U.K.

David Kritchevsky
The Wistar Institute ·
Philadelphia, Pennsylvania

Alan R. Liss, Inc., New York

Address all Inquiries to the Publisher
Alan R. Liss, Inc., 150 Fifth Avenue, New York, NY 10011

Copyright © 1983 Alan R. Liss, Inc.

Printed in the United States of America

Library of Congress Cataloging in Publication Data
Main entry under title:
Animal and vegetable proteins in lipid metabolism and atherosclerosis.

(Current topics in nutrition and disease; v. 8)
Bibliography: p.
Includes index.
1. Atherosclerosis—Nutritional aspects.   2. Proteins
—Physiological effect.   3. Lipids—Metabolism.
I. Kritchevsky, David, 1920–   . II. Gibney,
Michael J. III. Series. [DNLM: 1. Lipids—Metabolism.
2. Dietary proteins—Pharmacodynamics. 3. Vegetable
proteins—Pharmacodynamics. 4. Arteriosclerosis—Etiology.
5. Cholesterol—Blood. W1 CU82R v. 8 / QU 55 A598]
RC692.A644   1983   616.3'997071   82-23961
ISBN 0-8451-1607-X

Contents

Contributors

K.K. Carroll [9]
Department of Biochemistry, University of Western Ontario, London, Ontario N6A 5C1, Canada

Susanne K. Czarnecki [1,85]
The Wistar Institute of Anatomy and Biology, Philadelphia, Pennsylvania 19104; *currently*, National Heart, Lung and Blood Institute, National Institutes of Health, Bethesda, Maryland 20205

G.C. Descovich [135]
II Medical Clinic, University of Bologna, Italy

Patrick J. Gallagher [149]
Departments of Pathology and Nutrition, Faculty of Medicine, University of Southampton, Southampton, SO9 5NH England, U.K.

Michael J. Gibney [ix,149]
Departments of Pathology and Nutrition, Faculty of Medicine, University of Southampton, Southampton, SO9 5NH England, U.K.

R.J.J. Hermus [19]
Central Institute for Nutrition and Food Research (CIVO-TNO), 3700 AJ Zeist, The Netherlands

Martijn B. Katan [111]
Department of Human Nutrition, Agricultural University, 6703 BC Wageningen, The Netherlands

D.N. Kim [101]
Department of Pathology, Neil Hellman Building, Albany Medical College, Albany, New York 12208

David M. Klurfeld [85]
The Wistar Institute of Anatomy and Biology, Philadelphia, Pennsylvania 19104

David Kritchevsky [ix,1,85]
The Wistar Institute of Anatomy and Biology, Philadelphia, Pennsylvania 19104

K.T. Lee [101]
Department of Pathology, Neil Hellman Building, Albany Medical College, Albany, New York 12208

The number in brackets indicates the opening page of the contibutor's article.

G. Noseda [135]
Beata Vergine Hospital, Mendrisio, Switzerland

Joop M.A. van Raaij [111]
Department of Human Nutrition, Agricultural University, 6703 BC Wageningen, The Netherlands

J.M. Reiner [101]
Department of Pathology, Neil Hellman Building, Albany Medical College, Albany, New York 12208

C.R. Sirtori [135]
Center E. Grossi Paoletti, University of Milan, Milan, Italy

Jon A. Story [85]
The Wistar Institute of Anatomy and Biology, Philadelphia, Pennsylvania 19104; *currently,* Department of Foods and Nutrition, Purdue University, West Lafayette, Indiana 47907

Michihiro Sugano [51]
Laboratory of Nutrition Chemistry, Kyushu University School of Agriculture, Fukuoka 812, Japan

Shirley A. Tepper [85]
The Wistar Institute of Anatomy and Biology, Philadelphia, Pennsylvania 19104

A.H.M. Terpstra [19]
Department of Human Nutrition, Agricultural University, 6703 BC Wageningen, The Netherlands

W.A. Thomas [101]
Department of Pathology, Neil Hellman Building, Albany Medical College, Albany, New York 12208

Clive E. West [19,111]
Department of Human Nutrition, Agricultural University, 6703 BC Wageningen, The Netherlands

Preface

This volume is a result of a workshop on the subjects of animal and vegetable protein effects on lipid metabolism and atherosclerosis which was held during the XII International Congress of Nutrition in San Diego, California, 1981. The workshop was structured for maximum discussion and did not allow for the formal presentation of results by the large numbers of participants. Accordingly we felt that the many topics discussed deserved publication so that they could reach a wider audience. The essays in this volume cover the effects of animal and vegetable proteins and amino acid additions to these proteins. The chapters range from studies on lipid metabolism and atherosclerosis carried out in rats, rabbits, and swine to effects of proteins on human serum lipids and lipoproteins.

The first studies on experimental atherosclerosis were carried out because of interest in the effects of animal proteins. After a long hiatus, there is a renewed interest in the effects of protein *per se* on lipid metabolism and atherosclerosis. This volume is a "state-of-the-art" summary of current knowledge.

Michael J. Gibney
David Kritchevsky

Animal and Vegetable Proteins in Lipid
Metabolism and Atherosclerosis, pages 1–7
© *1983 Alan R. Liss, Inc., 150 Fifth Ave., New York, NY 10011*

1
Dietary Protein and Experimental Atherosclerosis: Early History

David Kritchevsky and Susanne K. Czarnecki
The Wistar Institute of Anatomy and Biology, 36th Street at Spruce, Philadelphia, Pennsylvania 19104

The first purely nutritional studies in experimental atherosclerosis were carried out by Ignatowski [1908a,b, 1909]. Prior to his work, the two major theories regarding the etiology of atherosclerosis —mechanical injury or toxicity — were tested by mechanical injury to the arteries, injection of noxious agents, or stress. Ignatowski was influenced by the work of Garnier and Simon [1907] who fed beef or horsemeat to rabbits and observed pathological changes in the livers and very poor survival. The liver changes were attributed to toxic acidic products of protein digestion. When Ignatowski [1908a] fed two adult rabbits a diet consisting only of meat (50–100 g/day), they lost weight and succumbed within ten days. Rabbits fed 0.6–15 g of meat daily for long periods (up to eight months) developed nephritis, liver cirrhosis, and pronounced atherosclerosis. In his further studies, Ignatowski [1908b, 1909] fed newborn rabbits combinations of milk and egg yolk and observed anemia, cirrhosis, renal disorders, and atherosclerosis.

Lubarsch [1909, 1910] fed rabbits 12 g per day of powdered liver, adrenal powder, or horsemeat. He observed aortic fatty streaks within two weeks in the animals fed liver; after 3–4 months, he observed calcified aortic lesions. Rabbits fed horsemeat showed no ill effects; those fed adrenal powder failed to survive. Lubarsch theorized that the diets injured arterial smooth muscle leading to intimal damage and collagen and elastin proliferation.

Susanne Czarnecki's present address is National Heart, Lung and Blood Institute, National Institutes of Health, Bethesda, Maryland 20205.

Stuckey [1910a,b, 1911] devised an experiment to test for the atherogenic component of animal protein. He fed five groups of rabbits the following diets daily: (1) milk plus egg albumen, (2) milk plus meat juice, (3) egg yolk in milk, (4) egg plus meat juice in milk, and (5) control. All groups received a specified amount of the diets, and oats and hay ad libitum. All four test groups exhibited aortic lesions with those in groups fed egg yolk (3 and 4) being most severe. He concluded that egg yolk was the primary atherogenic factor. The plaques did not regress when the rabbits were returned to a diet of vegetables and greens. Later, Stuckey [1912] compared the effects of tallow, cod liver oil, sunflower seed oil, and ox brain emulsified in milk using gum arabic. Lesions were observed only in the aortas of rabbits fed ox brain. Stuckey concluded that a nonprotein component common to egg yolk and ox brain was the atherogenic agent. On the basis of histological studies, Chalatow [1912] suggested that the birefringent material observed in the livers of rabbits fed by Stuckey was a mixture of cholesterol, lecithin, and fatty acids. Fahr [1912] fed two groups of rabbits a diet similar to that used by Ignatowski. After ten months of feeding, he observed moderate to severe atherosclerosis and elevated blood pressure in all the animals. Wesselkin [1912, 1913] carried out experiments to test the individual effects of these materials. He fed rabbits (1) egg yolk plus egg albumen, (2) egg albumen plus lecithin, (3) egg yolk, or (4) lecithin. All the test substances were fed dissolved in milk. Only rabbits fed egg yolk exhibited atherosclerosis and fatty livers. Wesselkin suggested that cholesterol, not lecithin, was the atherogenic factor.

In 1913, Anitschkow and Chalatow [1913] and Wacker and Hueck [1913a,b] independently and simultaneously produced atherosclerosis and fatty livers in rabbits by feeding cholesterol. Anitschkow and Chalatow fed 0.5–0.8 g cholesterol/day dissolved in vegetable oil for 1–2 months, whereas Wacker and Hueck fed 1.25 g cholesterol/day mixed with oats for 3–5 months. Anitschkow and Chalatow cited the earlier experiments using egg yolk and ox brain as the basis for their use of cholesterol: "The only fact that seems important to us is that pure cholesterol or cholesterol-containing substances bring about pathologic processes in the rabbit." Wacker and Hueck believed that hypercholesterolemia was a prerequisite condition for production of atherosclerosis, but pointed out that the severity of lesions did not correlate with either duration of feeding or the amount of cholesterol ingested. The rabbit maintained on the diet for the longest time had a normal aorta. These two studies set the stage for more elaborate research into the production of experimental lesions. They also eclipsed for decades research on atherogenic effects of other components of the diet.

Kon [1913, 1914] found that beef liver powder mixed with a soybean preparation was atherogenic but defatted beef liver was not. Horse liver, which contains one-twelfth of the ether-extractable material of beef liver, was not atherogenic. Kon suggested that the level of fat in the diet was important for production of aortic lesions. Kon and Yamada [1916] found that rabbits fed 5 g of lanolin daily exhibited aortic lesions similar to those seen in rabbits fed egg yolk, liver powder, or cholesterol.

Saltykow [1913, 1914] fed rabbits 400 g milk and 1 g bread for 1–2 years and produced moderate to severe atherosclerosis. The quantity of milk fed contained about 45–60 mg cholesterol and 14–15 g protein. Steinbisss [1913] and van Leersum [1912, 1914] fed rabbits horsemeat or horse liver with varying atherogenic effects and concluded that some abnormal metabolites, rather than the dietary components, were responsible for the observed effects. Van Leersum [1912] found that rabbits fed 5–12 g/day of horse liver exhibited elevated blood pressure. Knack [1915] questioned the type of diets used in the earlier studies and carried out an experiment in which one group of rabbits was fed a normal diet plus 0.5–4.5 g/day of cholesterol enclosed in pellets of moistened bread, whereas another group was fed one egg, milk, and greens daily. The latter diet provided about 300 mg of cholesterol/day. Most of the animals fed the normal diet plus cholesterol (8/12) were free of aortic lesions, whereas all of the second group exhibited well-defined atherosclerosis. Knack, too, attributed the atherosclerosis observed to abnormal metabolic products and suggested that factors other than dietary cholesterol and hypercholesterolemia were involved in the establishment of atherosclerosis.

Schmidtmann [1922] found that rabbits fed liver powder daily became hypertensive and atherosclerotic, whereas those fed the same amount of dried muscle had only slight increases in blood pressure and no lipid deposits in the aorta. He hypothesized that dietary cholesterol was the hypertensive agent. After comparing the effects of horse liver powder and cholesterol, Deicke [1926] concluded that there was no strong correlation between dietary cholesterol, hypercholesterolemia, and atherosclerosis.

Monckeberg [1924] reviewed the field on the twentieth anniversary of the coining of the term atherosclerosis. He questioned the significance for human disease of alteration of cholesterol metabolism and elevation of blood pressure, and wondered if the two conditions act independently or if lipid infiltration is an initiator. Schoenheimer [1924] reviewed experimental cholesterolemia in the rabbit. He favored the view that in the various regimens used cholesterol was the agent leading to atherosclerosis. He pointed out that herbivores retain cholesterol in the blood and organs, whereas omnivores excrete it.

While studying Bright's disease, Newburgh [1919] observed that rabbits fed 15–30 g casein daily developed atherosclerosis. Newburgh and Squier [1920] fed rabbits a mixture of powdered beef and flour (1:2) for 1–7 months and found the severity of atherosclerosis was roughly proportional to the duration of feeding. Further investigations were carried out using diets which contained dried beef, flour, and bran and provided 27% or 36% protein. They found that the time required for appearance of lesions and their severity were functions of protein level and duration of feeding. The animals fed 27% and 36% protein diets provided 28 and 36 mg cholesterol/day, an amount insufficient to be atherogenic per se. Newburgh and Clarkson [1922, 1923a,b] concluded that the observed lesions were due to the protein in the diet.

In an effort to determine the threshold at which dietary cholesterol became atherogenic, Clarkson and Newburgh [1926] fed rabbits a normal diet augmented by 25, 113, 253, or 507 mg of cholesterol daily (administered in capsules). Rabbits fed 25 mg of cholesterol daily for 51–288 days were normocholesterolemic and exhibited no lesions. Rabbits fed 113 mg of cholesterol daily for 12–302 days were normocholesterolemic and one of 19 had slight atherosclerosis. Of the rabbits fed 253 mg of cholesterol for 33–256 days, twelve survived for fewer than 125 days and only one was atherosclerotic; 8/13 rabbits living for more than 147 days developed aortic lesions and 4/13 were hypercholesterolemic. Five of seven rabbits fed 507 mg of cholesterol daily for 47–87 days exhibited a moderate degree of atherosclerosis. The overall conclusion from these experiments was that the amount of dietary cholesterol needed to produce atherosclerosis in rabbits was at least ten times the amount present in the atherogenic high-protein diets. Nuzum and his associates [1925, 1926, 1927] fed rabbits diets containing liver, casein, and cereal (41% protein), oats (16% protein) or soybeans (36% protein). Rabbits fed liver protein for 3–11 months exhibited severe atherosclerosis and hypertension; those with the highest blood pressure presented the most severe pathological changes. Rabbits fed the oat diet for 24 months exhibited atherosclerosis and hypertension; those fed soybeans for two years were hypertensive, but with normal aortas. Nuzum agreed with Newburgh that high-protein diets were, in themselves, sufficient to produce atherosclerosis and hypertension. Anderson [1926a,b] and Nuzum et al [1930, 1932] conducted other experiments relating long-term feeding of animal protein to kidney function and always observed atherosclerosis as a concomitant. Meeker and Kesten [1940, 1941] were the first investigators to compare the effects of animal and vegetable proteins and found that animal protein (casein) was more atherogenic than vegetable protein (soy).

As Bischoff [1932] has pointed out, the early studies were often nutritionally deficient and poorly detailed. However, with all their defects, they provided data which are remarkably similar to those being obtained today in laboratories which pay meticulous attention to nutritional details. The pioneering studies, whatever their faults, set the stage for today's exhaustive investigations into dietary influences on atherosclerosis. It has been pointed out [Kritchevsky, 1976, 1979] that all dietary components and their interactions play a role in atherosclerosis. The contribution of protein to the development and progression of atherosclerosis is gaining new recognition. The new work is building on the observation made over 70 years ago.

ACKNOWLEDGMENTS

This work was supported in part by grants HL-03299 and CA-09171, and a Research Career Award (HL-00734) from the National Institutes of Health. Also supported by grant 59-2426-0-1-479 from the US Department of Agriculture, SEA.

REFERENCES

Anderson H (1926a): Experimental renal insufficiency. The effects of high protein diet in the presence of low renal function on the kidneys, aorta and liver; changes in the blood pressure and concentration of blood metabolites. I. Controls on normal diets. Arch Intern Med 37: 297–312.

Anderson H (1926b): Experimental renal insufficiency. The effects of high protein diet in the presence of low renal function on the kidneys, aorta and liver; changes in the blood pressure and concentration of blood metabolites. II. Protein diet experiments. Arch Intern Med 37: 313–335.

Anitschkow N, Chalatow S (1913): Ueber experimentelle cholesertin-steatose and ihre bedeutung fur die entstehung einiger pathologischer prozess. Zentralbl Alleg Pathol Patholog Anat 24:1–9.

Bischoff F (1932): The influence of diet on renal and blood vessel changes. J Nutr 5:431–450.

Chalatow SS (1912): Uber das verhalten der leber gegenuber den verschiedenen arten von speise-fett. Virchows Arch Pathol Anat Physiol Klin Med 207:452–469.

Clarkson S, Newburgh LH (1926): The relation between atherosclerosis and ingested cholesterol in the rabbit. J Exp Med 43:595–612.

Deicke O (1926): Beobachtungen an kaninchen mit kunstlicher cholesterinzufuhr. Krankheits-forschung 3:399–418.

Fahr T (1912): Beitrage zur experimentellen atherosklerose unter besonderer beruchsichtigung der frage nach dem zusammenhang zwischen nebenierenveranderungen und atherosklerose. Verh Dtsch Pathol Ges 15:234–249.

Garnier M, Simon LG (1907): De l'etat due foie chez les lapins soumis au regime carne. Comptes Rendus Herbomadaires des Seances et Memoires de la Societe de Biologie 63:250–252.

Ignatowski A (1908a): Influence de la nourriture animale sur L'organisme des lapins. Arch Med Exp Anat Pathol 20:1–20.

Ignatowski A (1908b): Changes in parenchymatous organs and in the aorta of rabbits under the influence of animal protein. Izvestiya Imperatorskoi Voyenno-Meditsinskoi Akademii (St. Petersburg) 18:231–244.

Ignatowski A (1909): Uber die Wirbung des tierischen eiweisses auf die aorta und die paren-chymatosen organe der kaninchen. Virchows Arch Pathol Anat Physiol Klin Med 198: 248–270.

Knack AV (1915): Uber cholesterinsklerose. Virchows Arch Pathol Anat Physiol Klin Med 220:36–52.

Kon Y (1913): Referat uber arteriosklerose. Trans Jpn Pathol Soc 3:8–19.

Kon Y (1914): Futterungsversuche an saugetieren mit leberpulver und eigelb. Trans Jpn Pathol Soc 4:105–112.

Kon Y, Yamada H (1916): Weitere mitteilung uber die experimentellen studien der lipoidstoff-wechselstorungen. Trans Jpn Pathol Soc 6:98–101.

Kritchevsky D (1976): Diet and atherosclerosis. Am J Pathol 84:615–632.

Kritchevsky D (1979): Dietary interaction. In Levy RI, Rifkind BM, Dennis BH, Ernst ND (eds): "Nutrition, Lipids and Coronary Heart Disease. A Global View." New York: Raven Press, pp 229–246.

Lubarsch O (1909): Zur pathogenese der atherosklerose der arterien. Muenchener Med Wochenschr 56:1819–1820.

Lubarsch O (1910): Ueber alimentare schlagaderverbolkung. Muenchener Med Wochenschr 57: 1577–1580.

Meeker DR, Kesten HD (1940): Experimental atherosclerosis and high protein diets. Proc Soc Exp Biol Med 45:543–545.

Meeker DR, Kesten HD (1941): Effect of high protein diets on experimental atherosclerosis of rabbits. Arch Pathol 31:147–162.

Monckeberg JG (1924): Arteriosclerose. Klin Wochenschr 3:1473–1478.

Newburgh LH (1919): The production of Bright's disease by feeding high protein diets. Arch Intern Med 24:359–377.

Newburgh LH, Clarkson S (1922): Production of arteriosclerosis in rabbits by diets rich in animal protein. J Am Med Assoc 79:1106–1108.

Newburgh LH, Clarkson S (1923a): The production of arteriosclerosis in rabbits by feeding diets rich in meat. Arch Intern Med 31:653–676.

Newburgh LH, Clarkson S (1923b): Renal injury produced in rabbits by diets containing meat. Arch Intern Med 26:38–40.

Newburgh LH, Squier TL (1920): High protein diets and arteriosclerosis in rabbits: A preliminary report. Arch Intern Med 26:38–40.

Nuzum FR (1927): Changes in the kidney in animals with increased blood pressures while on high protein diets. Arch Intern Med 40:364–376.

Nuzum FR, Elliot AH, Priest BV (1932): Spontaneous nephritis in the rabbit. Arch Intern Med 49:744–752.

Nuzum FR, Osborne M, Sansum WD (1925): The experimental production of hypertension. Arch Intern Med 35:492–499.

Nuzum FR, Elliot AH, Evans RD, Priest BV (1930): The occurrence and nature of spontaneous arteriosclerosis and nephritis in the rabbit. Arch Pathol 10:697–716.

Nuzum FR, Seegal B, Garland R, Osborne M (1926): Arteriosclerosis and increased blood pressure. Arch Intern Med 37:733–744.

Saltykow S (1913): Zur kenntnis der alimentaren krankheiten der versuchstiere. Virchows Arch Pathol Anat Physiol Klin Med 213:8–17.

Saltykow S (1913–1914): Experimentelle atherosklerose. Beitr Patholog Anat Allg Pathol 57: 415–473.

Schmidtmann M (1922): Experimentelle studien zur pathogenese der arteriosklerose. Virchows Arch Pathol Anat Physiol Klin Med 237:1–21.

Schoenheimer R (1924): Uber die experimentelle cholesterinkrankiert der kaninchen. Virchows Arch Pathol Anat Physiol Klin Med 244:1–42.

Steinbiss W (1913): Uber experimentelle alimentare atherosklerose. Virchows Arch Pathol Anat Physiol Klin Med 212:152–187.

Stuckey NW (1910a): Alterations in the aorta of rabbits under the influence of excessive animal diet. Saint Petersburg, Russia: Imperial Military Medical Academy, Doctoral Dissertation #17.

Stuckey NW (1910b): Uber die veranderungen der kaninchen aorta bei der reichlichen tierischen kost. Zentralbl Allg Pathol Patholog Anat 21:668.

Stuckey NW (1911): Uber die veranderungen der kaninchen aorta unter der wirkung reichlicher tierscher nahrung. Zentralbl Allg Pathol Patholog Anat 22:379–380.

Stuckey NW (1912): Uber die veranderungen der kaninchen-aorta bei der futterung mit verschiedenen fettsorten. Zentralbl Allg Pathol Patholog Anat 23:910–911.

van Leersum EC (1912): Alimentare blutdruckerhohung. Z Exp Pathol Therap 1:408–425.

van Leersum EC (1914): Zur frage der experimentelen alimentaren atherosklerose. Virchows Arch Pathol Anat Physiol Klin Med 217:452–462.

Wacker L, Heuck W (1913a): Chemische and morphologische untersuchungen uber die bedeutung des cholesterins im organismus. Arch Exp Pathol Pharmakol 74:416–441.

Wacker L, Hueck W (1913b): Ueber experimentelle atherosklerose und cholesterinamie. Muenchener Med Wochenshr 60:2097–2100.

Wesselkin NW (1912): Concerning the deposition of fat-like substances in the organs. Russky Vrach 11:1651–1655.

Wesselkin NW (1913): Uber die Ablagerung von fettartigen stoffen in den organen. Virchows Arch Pathol Anat Physiol Klin Med 212:225–235.

Animal and Vegetable Proteins in Lipid
Metabolism and Atherosclerosis, pages 9–17
© 1983 Alan R. Liss, Inc., 150 Fifth Ave., New York, NY 10011

2

Dietary Proteins and Amino Acids — Their Effects on Cholesterol Metabolism

K.K. Carroll

Department of Biochemistry, University of Western Ontario, London, Ontario, N6A 5C1, Canada

I. INTRODUCTION

Our interest in the effects of dietary proteins and amino acids on cholesterol metabolism arose from our attempts to understand why rabbits become hypercholesterolemic and develop atherosclerosis when fed semipurified diets without added cholesterol, as reported by Lambert et al [1958] and Malmros and Wigand [1959]. By varying the composition of the semipurified diet, we were able to show that the effects were associated with the casein commonly used as the protein component of semipurified diets, and could be prevented by replacing the casein with soy protein [Hamilton and Carroll, 1976].

After one month, the plasma cholesterol levels of rabbits fed a low-fat, cholesterol-free, semipurified diet containing casein averaged 200 mg/dl, whereas those fed a similar diet containing soy protein isolate in place of the casein had

K.K. Carroll is a Career Investigator of the Medical Research Council of Canada.

plasma cholesterols averaging 70 mg/dl, which is about the level observed in rabbits fed diets formulated from natural ingredients. In longer-term experiments, the average plasma cholesterol levels of rabbits fed the cholesterol-free, casein diet were usually in the range of 200–300 mg/dl, while those of rabbits fed the corresponding soy protein diet remained below 100 mg/dl. After ten months on diet, the rabbits fed casein showed moderate to severe atherosclerosis, whereas the aortas of rabbits fed the soy protein had only minimal lesions [Carroll et al, 1979].

Feeding trials with semipurified diets containing various other delipidated proteins showed that in general the diets containing animal proteins tended to produce a hypercholesterolemia, whereas those containing plant proteins gave low plasma cholesterol levels (Fig. 2-I). These observations, and the original reports of hypercholesterolemia and atherosclerosis produced by cholesterol-free, semipurified diets, have stimulated work in a number of other laboratories, much of which is summarized in this volume. The following account will therefore be confined largely to studies carried out in our laboratory. The primary objectives of these studies were to determine which components of dietary proteins are responsible for their differing effects on plasma cholesterol levels and to elucidate the mechanisms by which dietary protein may influence these levels through effects on cholesterol metabolism.

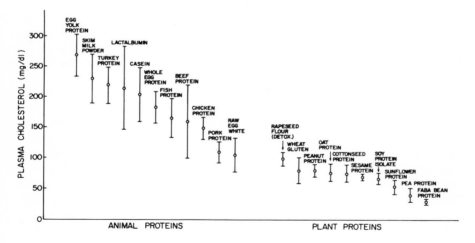

Fig. 2-I.   Effects on plasma cholesterol levels of feeding low-fat, cholesterol-free, semipurified diets containing different delipidated proteins. The diets were fed to groups of 5–6 rabbits for 28 days and the results are shown as mean ± SE (reproduced from Carroll and Huff [1980] by permission of the publisher).

II. FURTHER STUDIES ON EFFECTS OF DIETARY PROTEINS AND OTHER DIETARY COMPONENTS ON PLASMA CHOLESTEROL LEVELS IN RABBITS

The hypercholesterolemia produced by a cholesterol-free, semipurified diet containing casein was readily reversible, as shown by alternating diets containing casein or soy protein [Carroll and Huff, 1980]. However, the response in older rabbits was more sluggish than that in young, actively growing rabbits [Huff et al, 1982]. Doubling the amount of casein in the diet tended to enhance the hypercholesterolemia, whereas doubling the amount of soy protein had no effect on the plasma cholesterol level [Huff et al, 1977]. Terpstra et al [1981] also found that the degree of hypercholesterolemia increased with increasing amounts of casein in the diet.

A diet containing a 1:1 mixture of casein and soy protein produced no increase in plasma cholesterol, while a 3:1 mixture gave a level intermediate between those obtained with either casein or soy protein alone [Huff et al, 1977]. Kritchevsky et al [1981] reported that the hypercholesterolemic effect of beef protein was also largely prevented by feeding it as a 1:1 mixture with textured soy protein. In our experiments, rabbits fed the mixtures of casein and soy protein showed better weight gains than those fed either protein alone. The results of our other feeding trials described above also indicated that the effects of dietary protein on plasma cholesterol levels were not related to body-weight gain.

Low-fat diets containing 1% by weight of corn oil to provide essential fatty acids were used in all of the above experiments carried out in our laboratory. In one study, however, it was shown that casein and soy protein diets containing 15% by weight of butter gave results similar to those obtained with the corresponding low-fat diets [Carroll and Hamilton, 1975]. When 15% corn oil was incorporated into a casein diet, the hypercholesterolemia was largely prevented, in agreement with the original reports of Lambert et al [1958] and Malmros and Wigand [1959]. Incorporating 15% corn oil into a diet containing soy protein made little different in its effect on plasma cholesterol.

Other experiments showed that the hypercholesterolemic effect of dietary casein could be modified by the type of carbohydrate in the diet [Carroll and Hamilton, 1975; Hamilton and Carroll, 1976]. In particular, diets containing lactose, rice starch, or potato starch gave lower plasma cholesterol levels than diets containing glucose, sucrose, wheat starch, or corn starch. The results could also be modified by the type of fiber in the diet [Hamilton and Carroll, 1976; Carroll et al, 1978b]. Some fibrous materials, such as alfalfa or oat hulls, had a tendency to decrease the level of plasma cholesterol, whereas high levels of cellulose enhanced the hypercholesterolemic effect of dietary casein.

These results are in general agreement with the findings of others [Kritchevsky, 1979].

III. AMINO ACID COMPOSITION OF DIETARY PROTEINS IN RELATION TO THEIR EFFECTS ON PLASMA CHOLESTEROL LEVELS

It is of interest to know whether the differing effects of dietary proteins on plasma cholesterol levels are due to differences in their amino acid composition. We therefore investigated the effects of diets containing enzymatic digests of casein and soy protein, or mixtures of L-amino acids in the proportions found in these proteins [Huff et al, 1977]. The enzymatic digests had much the same effect on plasma cholesterol as the intact proteins. The mixture of amino acids corresponding to casein also produced a hypercholesterolemia similar to that obtained with casein itself. The mixture of amino acids corresponding to soy protein produced some elevation of plasma cholesterol and thus gave a rather different result than the intact protein.

Further studies with various mixtures of amino acids provided additional evidence that dietary amino acids can affect the level of plasma cholesterol in rabbits [Huff and Carroll, 1980a], but it was not clear from these studies which amino acids were most important in this regard. Kritchevsky [1979] suggested that the level of plasma cholesterol is positively correlated with the ratio of lysine to arginine in the diet, but analysis of the results of our feeding trials with mixtures of amino acids failed to provide strong evidence for this suggestion [Carroll, 1981a].

Examination of the amino acid composition of the various animal and plant proteins used in our earlier feeding trials showed that the animal proteins generally had higher levels of a number of essential amino acids than the plant proteins (Fig. 2-II). Arginine, on the other hand, tended to be higher in the plant proteins. The concentrations of nonessential amino acids varied over much the same ranges in both plant and animal proteins, with the exception of glutamic acid, which tended to be higher in plant proteins (Fig. 2-II). Preliminary studies in our laboratory have shown that supplementation of dietary soy protein with a mixture of essential amino acids has little or no effect on the plasma cholesterol level of rabbits. However, supplementation of a mixture of amino acids corresponding to soy protein with the same combination of essential amino acids produced a marked elevation in plasma cholesterol [Carroll

---

Fig. 2-II. Comparison of the amino acid composition of dietary proteins with the average plasma cholesterol levels obtained by feeding these proteins to groups of 5–6 rabbits for 28 days. The data on plasma cholesterol are for the same experiments as those of Figure 2-I. Reproduced from Carroll [1981b].

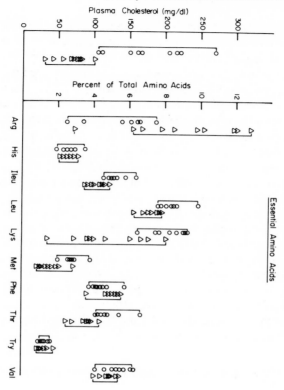

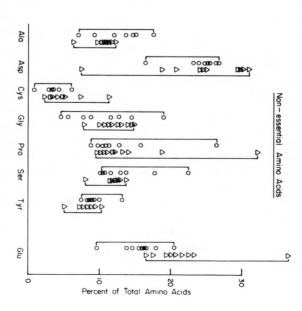

and Woodward, 1982]. Further experiments are being carried out to determine which of the essential amino acids are mainly responsible for this effect.

The effect on rabbit plasma cholesterol levels of feeding amino acid mixtures corresponding to egg yolk protein and sunflower seed protein was also investigated in our laboratory [Huff and Carroll, 1980a]. The mixture corresponding to egg yolk protein produced a hypercholesterolemia comparable to that obtained with the protein itself, while the mixture corresponding to sunflower seed protein gave a somewhat higher level than the intact protein. These results are analogous to those obtained with the amino acid mixtures corresponding to casein and soy protein described earlier.

The low levels of plasma cholesterol observed in animals fed plant proteins may thus be due in part to the presence of hypocholesterolemic substances in the plant protein preparations. An alternative and perhaps more plausible explanation is that the differences between the effects on plasma cholesterol levels of the plant proteins and their corresponding amino acid mixtures are related to factors involved in the digestion of the proteins and absorption of their constituent amino acids. It is unlikely that the absorption pattern can be duplicated by feeding a mixture of amino acids corresponding to the intact protein. Furthermore, it is known that a significant proportion of the amino acids derived from digestion of proteins are absorbed in the form of small peptides rather than individual amino acids [Matthews, 1975].

IV. STUDIES ON MECHANISM OF ACTION OF DIETARY PROTEIN

Other aspects of our work have been concerned with mechanisms by which dietary protein may affect cholesterol metabolism and levels of plasma cholesterol. Studies in which [26^{14}C]cholesterol was injected intravenously into rabbits fed cholesterol-free, semipurified diets showed that the cholesterol in plasma turned over more rapidly in rabbits fed soy protein than in those fed casein [Huff and Carroll, 1980b]. From the rate of expiration of $^{14}CO_2$ it was deduced that the rabbits fed soy protein also converted cholesterol to bile acids more rapidly than those fed casein. Furthermore, fecal analysis showed that the soy-fed animals excreted more neutral sterols and bile acids than casein-fed animals. The neutral sterol in the feces of casein-fed rabbits was mainly cholesterol, while the additional sterol excreted by the soy-fed rabbits was mainly coprostanol. The relative proportions of different bile acids was much the same in the feces of both groups, consisting of lithocholic, deoxycholic, and 12-keto-lithocholic acids in roughly equal proportions [Huff and Carroll, 1980b]. Studies on the absorption of cholesterol from the gut indicated that it was absorbed better by rabbits fed casein than by those fed soy protein [Huff and Carroll, 1980b]. Overall, it appears that the mechanisms for catabolism of cholesterol and for the excretion of cholesterol and its metabolites are more effective in rabbits fed soy protein than in those fed casein. This may be a reason for the lower plasma cholesterol levels in soy-fed rabbits.

Analysis of plasma lipoprotein patterns in rabbits fed casein or soy protein showed that the excess cholesterol in the plasma of casein-fed rabbits was carried in the low-density fractions, with little difference in the high-density lipoprotein [Carroll et al, 1979; Terpstra and Sanchez-Muniz, 1981]. Preliminary studies on the composition and turnover of the apoprotein components of plasma lipoproteins have also been carried out [Roberts et al, 1981]. As in the case of plasma cholesterol, the apoproteins of intermediate density lipoproteins in rabbits fed soy protein turned over more rapidly than those fed casein. The results with ^{125}I-labeled, very low density lipoproteins were less clear-cut but suggested the possibility that apoproteins of very low density lipoproteins from rabbits fed soy protein were transferred to high-density lipoproteins more readily than those from rabbits fed casein. The very low density lipoproteins of casein-fed rabbits appeared to contain a higher proportion of apo E than those from soy-fed rabbits [Roberts et al, 1981]. Further studies along these lines will no doubt help to determine how dietary protein influences the distribution and metabolism of cholesterol.

V. EFFECTS OF DIETARY PROTEIN ON PLASMA CHOLESTEROL LEVELS IN HUMANS

In conjunction with these studies in animals, we have also carried out a number of experiments to investigate the effects on plasma cholesterol of replacing animal protein by soy protein in human diets. The results have indicated that this causes some reduction in plasma cholesterol levels of both normal [Carroll et al, 1978a] and hypercholesterolemic subjects [Wolfe et al, 1981]. Much greater effects have been reported by Sirtori and his colleagues [Sirtori et al, 1977; Descovich et al, 1980], but others have reported little or no difference in plasma lipids between groups of subjects fed soy protein diets and groups fed diets containing mixtures of animal and plant protein [Holmes et al, 1980; Shorey et al, 1981; van Raaij et al, 1981]. It is evident that further work is required to determine the reasons for these differences.

ACKNOWLEDGMENTS

Support of these studies by the Ontario Heart Foundation is gratefully acknowledged.

VI. REFERENCES

Carroll KK (1981a): Soya protein and atherosclerosis. J Am Oil Chem Soc 58:416.

Carroll KK (1981b): Dietary protein and cardiovascular disease. In Bazán NG, Paoletti R, Iacono JM (eds): "Current Topics in Nutrition and Disease, Vol. 5. New Trends in Nutrition, Lipid Research, and Cardiovascular Diseases." New York: Alan R Liss, Inc., p 167.

Carroll KK, Giovannetti PM, Huff MW, Moase O, Roberts DCK, Wolfe BM (1978a): Hypocholesterolemic effect of substituting soybean protein for animal protein in the diet of healthy young women. Am J Clin Nutr 31:1312.

Carroll KK, Hamilton RMG (1975): Effects of dietary protein and carbohydrate on plasma cholesterol levels in relation to atherosclerosis. J Food Sci 40:18.

Carroll KK, Hamilton RMG, Huff MW, Falconer AD (1978b): Dietary fiber and cholesterol metabolism in rabbits and rats. Am J Clin Nutr 31:S203.

Carroll KK, Huff MW (1980): Dietary protein and cardiovascular diseases: Effects of dietary protein on plasma cholesterol levels and cholesterol metabolism. In Santos W, Lopes N, Barbosa JJ, Chaves D, Valente JC (eds): "Nutrition and Food Science: Present Knowledge and Utilization, Vol. 3. Nutritional Biochemistry and Pathology." New York: Plenum Publishing Corp., p 379.

Carroll KK, Huff MW, Roberts DCK (1979): Vegetable protein and lipid metabolism. In Wilcke HL, Hopkins DT, Waggle DH (eds): "Soy Protein and Human Nutrition." New York: Academic Press, p 261.

Carroll KK, Woodward CJH (1982): Hypocholesterolemic effects of soy protein in relation to amino acid composition: Experimental researches. In "Proceedings of the 4th International Meeting on Atherosclerosis: Etiopathogenesis, Clinical Evaluation and Therapy." Lancaster, England: MTP Press, Ltd (in press).

Descovich GC, Gaddi A, Mannino G, Cattin L, Senin U, Caruzzo C, Fragiacomo C, Sirtori M, Ceredi C, Benassi MS, Colombo L, Fontana G, Mannarino E, Bertelli E, Noseda G, Sirtori CR (1980): Multicentre study of soybean protein diet for outpatient hypercholesterolaemic patients. Lancet 2:709.

Hamilton RMG, Carroll KK (1976): Plasma cholesterol levels in rabbits fed low fat, low cholesterol diets. Effects of dietary proteins, carbohydrates and fibre from different sources. Atherosclerosis 24:47.

Holmes WL, Rubel GB, Hood SS (1980): Comparison of the effect of dietary meat versus dietary soybean protein on plasma lipids of hyperlipidemic individuals. Atherosclerosis 36:379.

Huff MW, Carroll KK (1980a): Effects of dietary proteins and amino acid mixtures on plasma cholesterol levels in rabbits. J Nutr 110:1676.

Huff MW, Carroll KK (1980b): Effects of dietary protein on turnover, oxidation, and absorption of cholesterol, and on steroid excretion in rabbits. J Lipid Res 21:546.

Huff MW, Hamilton RMG, Carroll KK (1977): Plasma cholesterol levels in rabbits fed low fat, cholesterol-free, semipurified diets: Effects of dietary proteins, protein hydrolysates and amino acid mixtures. Atherosclerosis 28:187.

Huff MW, Roberts DCK, Carroll KK (1982): Long-term effects of semipurified diets containing casein or soy protein isolate on atherosclerosis and plasma lipoproteins in rabbits. Atherosclerosis 41:327.

Kritchevsky D (1979): Vegetable protein and atherosclerosis. J Am Oil Chem Soc 56:135.

Kritchevsky D, Tepper SA, Czarnecki SK, Klurfeld DM, Story JA (1981): Experimental atherosclerosis in rabbits fed cholesterol-free diets. Part 9. Beef protein and textured vegetable protein. Atherosclerosis 39:169.

Lambert GF, Miller JP, Olsen RT, Frost DV (1958): Hypercholesteremia and atherosclerosis induced in rabbits by purified high fat rations devoid of cholesterol. Proc Soc Exp Biol Med 97:544.

Malmros H, Wigand G (1959): Atherosclerosis and deficiency of essential fatty acids. Lancet 2:749.

Matthews DM (1975): Intestinal absorption of peptides. Phyiol Rev 55:537.

Raaij JMA van, Katan MB, Hautvast JGAJ, Hermus RJJ (1981): Effects of casein versus soy protein diets on serum cholesterol and lipoproteins in young healthy volunteers. Am J Clin Nutr 34:1261.

Roberts DCK, Stalmach ME, Khalil MW, Hutchinson JC, Carroll KK (1981): Effects of dietary protein on composition and turnover of apoproteins in plasma lipoproteins of rabbits. Can J Biochem 59:642.

Shorey RL, Bazan B, Lo GS, Steinke FH (1981): Determinants of hypocholesterolemic response to soy and animal protein-based diets. Am J Clin Nutr 34:1769.

Sirtori CR, Agradi E, Conti F, Mantero O, Gatti E (1977): Soybean-protein diet in the treatment of type-II hyperlipoproteinaemia. Lancet 1:275.

Terpstra AHM, Harkes L, van der Veen FH (1981): The effect of different proportions of casein in semipurified diets on the concentration of serum cholesterol and the lipoprotein composition in rabbits. Lipids 16:114.

Terpstra AHM, Sanchez-Muniz FJ (1981): Time course of the development of hypercholesterolemia in rabbits fed semipurified diets containing casein or soybean protein. Atherosclerosis 39:217.

Wolfe BM, Giovannetti PM, Cheng DCH, Roberts DCK, Carroll KK (1981): Hypolipidemic effect of substituting soybean protein isolate for all meat and dairy protein in the diets of hypercholesterolemic men. Nutr Rept Int 24:1187.

Animal and Vegetable Proteins in Lipid
Metabolism and Atherosclerosis, pages 19–49
© *1983 Alan R. Liss, Inc., 150 Fifth Ave., New York, NY 10011*

3

Dietary Protein and Cholesterol Metabolism in Rabbits and Rats

A.H.M. Terpstra, R.J.J. Hermus, and C.E. West
Department of Human Nutrition, Agricultural University, De Dreijen 12,
6703 BC Wageningen (A.H.M.T., C.E.W.) and Central Institute for Nutrition
and Food Research (CIVO-TNO), PO Box 360, 3700 AJ Zeist (R.J.J.H.),
The Netherlands

I. INTRODUCTION

The role of diet and nutrition in the etiology of hypercholesterolemia and atherosclerosis was recognized by Ignatowski as far back as 1909. Serendipitously, he found that atherosclerosis could be induced by dietary means. Rabbits fed on diets containing meat, milk, and eggs very soon developed arterial lesions and the resemblance of these lesions to human atherosclerosis was immediately recognized. Initially, these arterial lesions were ascribed to the protein in the diet. However, this idea was largely abandoned when Anitschkow and Chalatow [1913] and Wacker and Hueck [1913] independently demonstrated that the feeding of crystalline cholesterol to rabbits also resulted in atherosclerosis. This prompted the idea that in the studies of Ignatowski [1909] the cholesterol in the diet rather than the protein source derived from animal products was the culprit. Nevertheless, a number of investigators over the years have clearly shown that in experimental animals dietary protein can play an important role in the genesis of hypercholesterolemia and atherosclerosis [Newburgh, 1919; Meeker and Kesten, 1941; Kritchevsky, 1964; Carroll and Hamilton, 1975].

Furthermore, epidemiological studies have provided evidence for a link between mortality from coronary heart disease (CHD) and the intake of animal protein, which is even better than that between CHD mortality and the intake of total or saturated fat [Yerushalmy and Hilleboe, 1957; Connor and Connor, 1972; Stamler, 1979]. However, it should be taken into account that dietary variables positively related to the mortality from CHD can be highly interrelated [Stamler, 1979]. Therefore, it is very difficult to find out which statistically significant correlations are etiologically significant and which are not.

Intervention studies in man have also shown that dietary protein might play an important role in the regulation of the levels of serum cholesterol. When animal protein in the diet was replaced by soy protein, a marked cholesterol-lowering effect was achieved. However, such results could be observed only when the subjects were hypercholesterolemic [Sirtori et al, 1979; Descovich et al, 1980; Wolfe et al, 1981]. In normocholesterolemic individuals, the type of dietary protein has been found to affect serum cholesterol only to a minor extent [Carroll et al, 1978] or not at all [van Raaij et al, 1981, 1982].

An overview of data from the literature concerning the role of dietary protein in cholesterol metabolism has been published previously [Terpstra, 1981; Terpstra et al, 1982a]. In this chapter, we present the results of studies from our laboratory on the effect of dietary protein on the concentration of serum cholesterol and lipoproteins in rabbits and rats.

II. DIETARY PROTEIN AND TOTAL SERUM CHOLESTEROL

A. Nature of Dietary Protein and Amino Acids

1. Rabbits. While studying the effect of various dietary fats on the level of serum cholesterol in rabbits, Hermus [1975] observed that the general health status and growth of the animals fed semipurified diets containing casein as the protein source were not satisfactory. Therefore, the casein in the diet was replaced by a protein mixture composed of fish protein, gelatin, casein, and soybean protein, which resembled more or less the amino acid composition of a commercial diet. This resulted in better growth of the rabbits together with lower levels of serum cholesterol compared with the animals fed on casein. Thus, there findings did confirm that the nature of the protein source in the diets of rabbits played an imporant role in determining the levels of serum cholesterol. Further, it was found that the growth of the rabbits on the semipurified diets could be improved by including bicarbonate in the diets. This prevented the development of acidosis which resulted in much better growth, irrespective of the protein source in the diet [Beynen and Wanrooij-Stroeken, 1981].

In a subsequent study, the cholesterolemic effects of various protein mixtures of increasing complexity were compared. Figure 3-I shows that substitution of these protein mixtures in the diet for casein brought about a considerable reduction in the levels of serum cholesterol.

In order to elucidate a possible role of the amino acid composition of the protein source, the effect of amino acid supplementation was investigated. A diet containing 12% casein was supplemented with amino acids to give the diet an individual amino acid content identical with a diet containing 20% protein

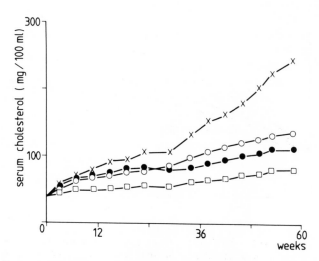

Fig. 3-I.   Concentration of serum cholesterol in rabbits fed semipurified diets containing various proteins: × —— ×, 20.8% casein (6 rabbits); ○ —— ○, 12% casein + 8% gelatin (8 rabbits); ● —— ●, 7.5% casein + 5% gelatin + 7.5% fish protein (9 rabbits); □ —— □, 6.2% casein + 4.3% gelatin + 6.2% fish protein + 4.1% soybean protein (9 rabbits) (from Hermus [1975] and Hermus et al [1979]).

derived from a mixture of casein, gelatin, and fish protein. In order to exclude a possible effect of feeding amino acids instead of intact proteins, a diet was also provided which contained a similar amount of casein, but supplemented with amino acids proportional to the composition of casein. All the "imitation proteins" induced higher levels of serum cholesterol than the intact proteins (Fig. 3-II). Nevertheless, the diet containing an amino acid composition resembling casein again resulted in higher levels of serum cholesterol than the diet with an amino acid composition resembling the protein mixture.

In the following study, casein was supplemented with arginine, alanine, and glycine. These amino acids are relatively scarce in casein compared with the protein mixture composed of casein, gelatin, and fish protein. Another diet was prepared in which amino acids abundantly present in casein were added to the protein mixture, so that the final amino acid composition was identical with that of the casein diet supplemented with arginine, alanine, and glycine. Feeding of these "imitation proteins" with an identical amino acid composition resulted in similar levels of serum cholesterol which were lower than those in rabbits fed a casein diet but higher than those in the animals fed the protein mixture (Fig. 3-III). Therefore, it was concluded that a considerable part of

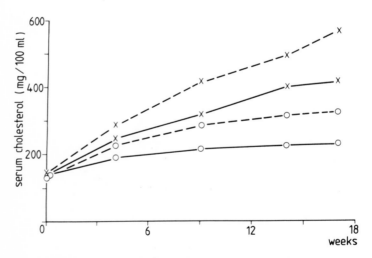

Fig. 3-II.   Concentration of serum cholesterol in rabbits fed semipurified diets containing various proteins and amino acid supplements (44.05): × —— ×, 20.8% casein (22 rabbits);
—— ○, 8% casein + 5.5% gelatin + 8% fish protein (14 rabbits); ○----○, 14% casein +
amino acids to give the diet an amino acid composition identical to the diet containing 8%
+ 5.5% gelatin + 8% fish protein (14 rabbits); ×----×, 14% casein + 7.2% amino
give the diet an amino acid composition identical to the diet containing 20.8% casein (14
from Hermus and Stasse-Wolthuis [1978] and Hermus et al [1979]).

the hypercholesterolemic effect of casein could be ascribed both to a relative shortage of some amino acids and to a relative excess of other amino acids. Further studies revealed that the addition of only glycine to a casein diet resulted in a considerable reduction in serum cholesterol levels, whereas supplementation with methionine had the opposite effect (Table 3-I). Subsequently, it was examined whether the hypocholesterolemic affect of glycine could be enhanced by gradually increasing the proportion of this amino acid in the diet. However, from Figure 3-IV it can be clearly seen that glycine supplementation did not show a dose-response relationship.

In another study, the role of the ratio of arginine to lysine in the diet in determining the level of serum cholesterol was examined. Casein is rich in lysine, but relatively poor in arginine, thus providing a very high lysine:arginine ratio. Lysine was added to a protein mixture composed of casein, fish protein, and

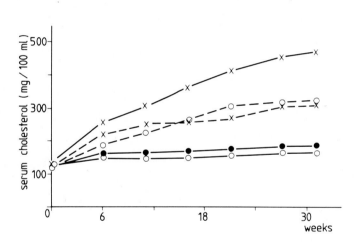

Fig. 3-III.   Concentration of serum cholesterol in rabbits fed semipurified diets containing various proteins and amino acid supplementations: × ——— ×, 20.8% casein (15 rabbits); ○ ——— ○, 8% casein + 5.6% gelatin + 8% fish protein (15 rabbits); ○----○, 20.8% casein supplemented with 0.34% arginine, 0.54% alanine, and 1.38% glycine to give the diet a concentration of these three amino acids equal to that of the diet containing 8% casein + 5.6% gelatin + 8% fish protein (15 rabbits); ×----×, 8% casein + 5.6% gelatin + 8% fish protein supplemented with 0.12% methionine, 0.29% lysine, 0.06% tryptophane, 0.48% leucine, 0.33% isoleucine, 0.44% tyrosine, 0.11% threonine, 0.42% valine, 0.09% histidine, 0.44% proline, 0.27% serine, 1.41% glutamine, and 0.24% phenylalanine to give the diet an amino acid composition equal to that of the diet containing 20.8% casein supplemented with 0.34% arginine, 0.54% alanine, and 1.38% glycine (15 rabbits); ● ——— ●, 8% casein + 5.6% gelatin + 8% fish protein supplemented with 1% lysine to give the diet a lysine/arginine ratio equal to that in casein (17 rabbits) (from Hermus et al [1979]).

gelatin in order to achieve a ratio of arginine to lysine equal to that of casein. However, no change in serum cholesterol was observed (Fig. 3-III). Similarly, the addition of arginine to a casein diet did not significantly affect the concentration of cholesterol in the serum (Table 3-I). These findings are in agreement with data obtained in rats by Nagata et al [1981]. They observed that addition of lysine to a diet containing soybean protein also failed to affect the serum cholesterol levels. However, contrasting results have been reported by Kritchevsky [1979]. In his studies with rabbits it was found that the addition of arginine to a casein diet resulted in a decrease in serum cholesterol levels, whereas lysine supplementation of a diet containing soybean protein had the reverse effect.

Further evidence for a role of the amino acid composition in determining the cholesterolemic property of dietary protein is provided by studies in which the intact protein was replaced by a mixture of amino acids. Table 3-I shows that feeding a mixture of amino acids resembling the amino acid composition of casein, but supplemented with glycine and arginine, resulted in serum cholesterol levels similar to those induced by intact casein also supplemented with

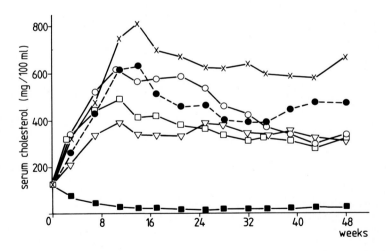

Fig. 3-IV. Concentration of serum cholesterol in rabbits fed semipurified diets containing casein and supplemented with various amino acids: × —— ×, 20.8% casein (13 rabbits); ●----●, 20.8% casein + 0.35% arginine + 0.54% alanine (13 rabbits); ∇ —— ∇, 20.8% casein + 0.35% arginine + 0.54% alanine + 0.65% glycine (15 rabbits); ○ —— ○, 20.8% casein + 0.35% arginine + 0.54% alanine + 1.40% glycine (15 rabbits); □ —— □, 20.8% casein + 0.35% arginine + 0.54% alanine + 3.15% glycine (15 rabbits); ■ —— ■, commercial diet (7 rabbits) (from Katan et al [1982]).

glycine and arginine. However, a mixture of amino acids with a composition identical with that of soybean protein induced serum cholesterol levels somewhat higher than those produced by feeding the intact soybean protein. Similar findings have been reported by Huff and Carroll [1980a], who observed that the feeding of intact casein and a mixture of amino acids resembling casein produced similar levels of serum cholesterol. A mixture of amino acids resembling soybean protein, however, was not able to maintain serum cholesterol levels as low as those achieved in the rabbits fed the intact soybean protein. Further studies by Huff and Carroll [1980a] in which amino acids were added to casein and soybean protein in order to provide an amino acid composition equivalent to soybean protein and casein, respectively, suggested to them that the intact protein component of each mixture had an overriding effect. Thus, all these studies in rabbits with amino acid supplementation revealed that the amino acid composition of a protein plays an important role in

TABLE 3-I. Effect of Dietary Protein and Amino Acid Supplementation on the Concentration of Serum Cholesterol in Rabbits

Diet[a]	Number of animals	Serum cholesterol concentration, mg/100 ml (mean ± SEM)
Experiment I		
Casein (20.8%)	22	477 ± 59
Casein (20.8%) + methionine (0.20%)	14	710 ± 88
Casein (20.8%) + arginine (0.80%)	14	548 ± 78
Casein (20.8%) + methionine (0.20%) + arginine (0.80%)	13	800 ± 117
Experiment II		
Casein (20.8%)	14	244 ± 55
Casein (20.8%) + glycine (1.38%)	10	106 ± 15
Casein (20.8%) + glycine (1.38%) + arginine (0.35%)	14	171 ± 33
Amino acids (to give diet an amino acid composition equal to that containing 20.8% casein + 1.38% glycine + 0.35% arginine)	15	159 ± 32
Soy protein (20.2%)	14	95 ± 11
Amino acids (to give diet an amino acid composition equal to that containing 20.2% soy protein)	14	155 ± 38
Casein (8%) + gelatin (5.6%) + fish protein (8%)	10	135 ± 36

[a]The diets in experiments I and II were fed for a period of 17 and 13 weeks, respectively. From Hermus and Dallinga-Thie [1979].

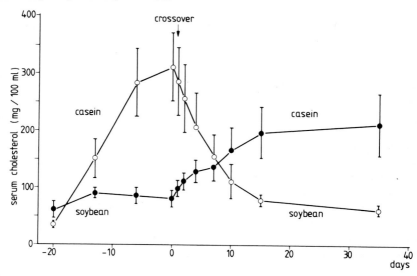

Fig. 3-VI. Concentration of serum cholesterol in rabbits fed semipurified diets containing either casein or soybean protein: ○ —— ○, casein before the crossover and soybean protein after the crossover; ● —— ●, soybean protein and casein, respectively. Each point denotes the mean from 6 rabbits; the vertical bars represent the SEM at each time point (from Terpstra et al [1982b]).

TABLE 3-III. Concentration of Serum and Liver Cholesterol in Lean Zucker Rats Fed Cholesterol-Enriched, Semipurified Diets Containing Different Proportions of Either Casein or Soybean Protein*

	Soybean protein		Casein	
	50%	20%	20%	50%
Males				
Serum cholesterol (mg/100 ml)				
Initial	94 ± 6	101 ± 4	104 ± 5	103 ± 2
Final	65 ± 3[a]	79 ± 4[a]	85 ± 4[a]	87 ± 6[a]
Liver cholesterol (g/100 g wet weight)	0.60 ± 0.03[a]	0.87 ± 0.07[b]	2.90 ± 0.31[c]	4.07 ± 0.43[c]
Females				
Serum cholesterol (mg/100 ml)				
Initial	105 ± 4	108 ± 6	109 ± 4	106 ± 7
Final	63 ± 4[a]	75 ± 3[a]	165 ± 20[b]	737 ± 52[a]
Liver cholesterol (g/100 g wet weight)	0.40 ± 0.02[a]	0.50 ± 0.03[b]	2.92 ± 0.43[c]	6.48 ± 0.45[d]

*Results are expressed as mean ± SEM, eight animals per group. The diets were fed for a period of 14 weeks.
Statistical comparison by a modified two-tailed t-test [Snedecor and Cochran, 1967]: horizontal values not sharing a common superscript are significantly different ($P < 0.05$).
From Terpstra et al [1982c].

ever, such effects were not observed, indicating a sex difference in susceptibility to the induction of changes in serum cholesterol levels by dietary means.
In both sexes, the concentration of cholesterol in the liver was influenced by
both the type and proportion of dietary protein. Rats fed on casein diets exhibited significantly higher levels of liver cholesterol than those fed soybean
protein. Furthermore, the increasing and lowering effect of casein and soybean protein, respectively, on liver cholesterol was enhanced by increasing the
proportion of the protein in the diet.

An elevation of the proportion of soybean protein in the diet of the female
rats did not lower the serum cholesterol levels further. Similar findings have
been reported in rabbits [Huff et al, 1979]. However, other studies with rats
[Moyer et al, 1956; Nath et al, 1959] have shown that the serum cholesterol-
lowering effect of dietary soybean protein could be enhanced by increasing the
proportion of the protein in the diet. It is possible that the serum cholesterol
levels of the female rats fed the diet containing 20% soybean protein were already so low that no further decrease could be achieved. More or less similar

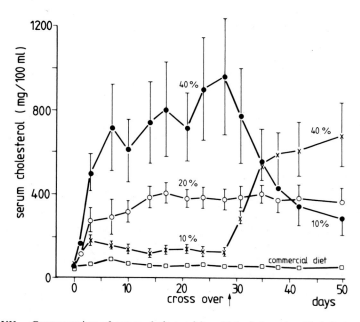

Fig. 3-VII. Concentration of serum cholesterol in rabbits fed semipurified diets containing
different proportions of casein: * —— *, 10% casein before the crossover and 40% casein after
the crossover (6 rabbits); ● —— ●, 40% and 10%, respectively (6 rabbits); ○ —— ○, 20%
throughout the whole experiment (6 rabbits); □ —— □, commercial diet (3 rabbits). The vertical bars represent the SEM at each time point (from Terpstra et al [1981a]).

findings have been reported in man. In hypercholesterolemic patients, a marked lowering of serum cholesterol could be achieved with dietary soybean protein [Sirtori et al, 1979], whereas in normocholesterolemic subjects no such effect was observed [van Raaij et al, 1981, 1982]. In order to test whether the cholesterol-lowering property of dietary soybean protein could be enhanced by the incorporation of a higher proportion of protein, the following experiment was conducted: Rats were rendered hypercholesterolemic by feeding them a diet containing 1.2% cholesterol and 30% casein. Subsequently, increasing amounts of soybean protein were included in this diet. Table 3-IV clearly shows that the incorporation of an additional amount of soybean protein in an atherogenic diet containing cholesterol and casein resulted in lower levels of serum cholesterol. This effect was enhanced by increasing the amount of additional soybean protein. Conversely, substitution of casein for the added soybean protein brought about an elevation of the serum cholesterol levels. Thus, this study clearly shows that dietary soybean protein is able to reduce the levels of serum cholesterol, whereas casein has the opposite effect; moreover, the hypo- and hypercholesterolemic property of soybean protein and casein, respectively, is enhanced by the incorporation of higher proportions of these proteins in the diet. In this context it is interesting to mention the studies of Kim et al [1978]. In swine fed high-cholesterol diets they observed that feeding a mixture of casein and soybean protein containing equal amounts of these two proteins resulted in serum cholesterol levels intermediate between those induced by the casein and soybean protein diets. Doubling the proportion of this protein mixture in the diet produced no further effects. This might indicate that upon increasing the proportion of this "neutral" protein mixture, the hypercholesterolemic action of casein and the hypercholesterolemic action of soybean protein were both enhanced to the same extent and therefore balanced each other.

C. Age and Cholesterolemic Response to Casein Diets

All foregoing experiments with rabbits have been performed in young, growing animals. However, in his paper in 1909, Ignatowski remarked that young rabbits were more susceptible to diet-induced atherosclerosis than their adult counterparts. Therefore, studies were carried out to examine the role of age in the development of hypercholesterolemia by feeding casein. Young and mature rabbits were fed semipurified diets containing either casein or soybean protein for a period of ten weeks. The hypercholesterolemic response produced by feeding a casein diet was much greater in the young rabbits than in the mature animals (Fig. 3-VIII). Subsequently, all the rabbits were switched back to the commercial diet and again transferred to the semipurified diets. During the second experimental period the increase in the levels of serum cholesterol in the "young" rabbits was less than during the first experimental

TABLE 3-IV. Concentration of Serum and Liver Cholesterol in Lean Female Zucker Rats Fed Cholesterol-Enriched, Semipurified Diets Containing Different Proportions of Casein and Soybean Protein*

	60% casein	45% casein	30% casein	30% casein plus 15% soybean protein	30% casein plus 30% soybean protein
Number of animals	12	9	11	12	12
Serum cholesterol (mg/100 ml)					
Initial	90 ± 3	90 ± 4	91 ± 3	90 ± 3	89 ± 3
5 weeks	339 ± 19[a]	201 ± 25[b]	101 ± 7[c]	84 ± 5[c,d]	82 ± 5[d]
11 weeks	372 ± 30[a]	200 ± 36[b]	109 ± 16[c]	81 ± 5[c]	64 ± 4[d]
18 weeks	288 ± 21[a]	218 ± 34[a]	120 ± 17[b]	92 ± 10[b]	66 ± 4[c]
Liver cholesterol (g/100 g wet weight)	6.06 ± 0.32[a]	3.36 ± 0.23[b]	2.62 ± 0.31[b]	1.37 ± 0.17[c]	0.80 ± 0.08[d]

*Results are expressed as mean ± SEM. Statistical comparison by a modified two-tailed t-test [Snedecor and Cochran, 1967]: horizontal values not having a common superscript are significantly different ($P < 0.05$). From Terpstra et al [1982d].

period. This might indicate that the young rabbits during the second experimental period were less susceptible to induction of hypercholesterolemia, as they had become older. Again during a third experimental period, the elevation in serum cholesterol in the casein-fed young rabbits was less than during the second period. Thus, these and similar findings reported by Huff et al [1982] indicate that the effect of diet on serum cholesterol metabolism of rabbits can be largely modulated by the age of the probands. It is interesting to note that similar findings have been reported in man. Descovich et al [1980] found that the hypocholesterolemic effect of dietary soybean protein was more pronounced in young than in older hypercholesterolemic patients.

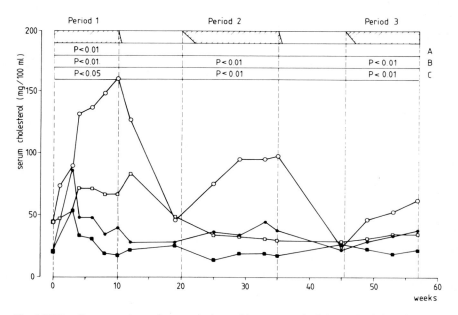

Fig. 3-VIII. Concentrations of serum cholesterol in young and adult rabbits fed semipurified diets containing either casein or soybean protein. The rabbits received the semipurified diets for three periods (indicated by the hatched area in the bar at the top of the figure). In the intermediate periods, a commercial diet was fed. ○ —— ○, young rabbits fed casein (12 animals); □ —— □, young rabbits fed soybean protein (12 animals); ● —— ●, adult rabbits fed casein (12 animals); ■ —— ■, adult rabbits fed soy protein (12 animals). Statistical comparison, using a two-tailed Wilcoxon test [Snedecor and Cochran, 1967]: between young and adult rabbits fed casein (A); between young rabbits fed casein and young rabbits fed soybean protein (B); and between adult rabbits fed casein and adult rabbits fed soybean protein (C) (from West et al [1982]).

D. Fat and Fiber in the Diet and Cholesterolemic Response to Dietary Casein

As it is known that dietary fat [Grundy, 1979] and fiber [Kritchevsky et al, 1977; Zilversmit, 1979; Stasse-Wolthuis, 1980] can affect cholesterol metabolism, a number of studies were carried out to examine their influence on casein-induced hypercholesterolemia. In order to study the influence of various types of dietary fiber, rabbits were first made hypercholesterolemic by feeding a semipurified diet containing casein and 21% sawdust for a period of four months. Subsequently, the animals were transferred to diets in which 50% of the sawdust was replaced by wheat bran, cellulose, or pectin. Table 3-V shows that the incorporation of pectin in the diet had a cholesterol-lowering tendency, whereas no such results were observed with wheat bran and cellulose. In another experiment, the type of fat in the diet was varied. When coconut oil in the diet was replaced by corn oil, the rabbits fed a semipurified casein diet exhibited levels of serum cholesterol comparable to those on a commercial diet (Table 3-VI). Thus, these studies show that in rabbits the hypercholesterolemic response to casein diets can be greatly modulated by the type of fat and fiber in the diet.

III. DIETARY PROTEIN AND SERUM LIPOPROTEINS

A. Rabbits

Cholesterol in the blood is transported in the form of lipoproteins, which can be subdivided into three major classes: the very low density lipoproteins (VLDL), the low-density lipoproteins (LDL), and the high-density lipoproteins (HDL). By means of density-gradient ultracentrifugation, these serum

TABLE 3-V. Effect of Various Sources of Dietary Fiber on the Concentration of Serum Cholesterol in Rabbits Fed Semipurified Diets Containing Casein

Diet[a]	Number of animals	Serum cholesterol (mg/100 ml ± SEM)			
		Initial	1 week	3 weeks	5 weeks
Sawdust	14	372 ± 67	457 ± 91	384 ± 68	396 ± 76
Wheat bran	14	370 ± 64	378 ± 69	365 ± 82	334 ± 72
Cellulose	14	371 ± 69	394 ± 73	380 ± 61	360 ± 60
Pectin	13	335 ± 53	237 ± 42	217 ± 42	224 ± 40
Commercial diet	4	36 ± 10	23 ± 6	21 ± 5	26 ± 9

[a]All the animals, with the exception of those on the commercial diet, had been fed a semipurified casein diet containing 21% sawdust for a period of 4 months. Subsequently, 50% of the sawdust in the diet was replaced by wheat bran, cellulose and pectin, respectively.

From Van Vliet PW, Nagelsmit A, Martinez de Prado M, Hermus RJJ, Katan MB (unpublished observations).

lipoproteins can be easily separated from each other [Terpstra et al, 1981b]. Figure 3-IX shows the density profile of the prestained serum lipoproteins of a normocholesterolemic rabbit compared with that of a human and various other animal species. In contrast to man, the rabbit has a pronounced HDL band and a faint LDL band. These visual impressions are reflected in the cholesterol concentrations in these lipoprotein fractions. In man, most of the serum cholesterol is transported in the LDL fraction, whereas in normocholesterolemic rabbits the HDL fraction is the main carrier of cholesterol. However, in an experiment in which rabbits were transferred from a commercial diet to a

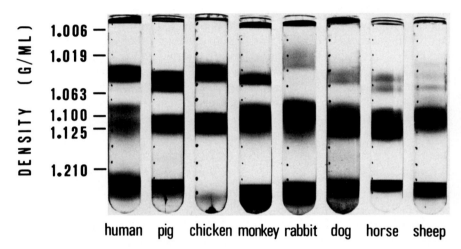

Fig. 3-IX. Photograph of the density profile of prestained serum lipoproteins from human and from various animal species observed after density-gradient ultracentrifugation (from Terpstra et al [1982d]).

TABLE 3-VI. Effect of Corn and Coconut Oil on the Concentration of Serum Cholesterol in Rabbits Fed Semipurified Diets Containing Casein

Diet	Number of animals	Serum cholesterol (mg/100 ml ± SEM)			
		Initial	1 week	2 weeks	3 weeks
Coconut oil	8	50 ± 5	174 ± 11**	203 ± 32*	208 ± 28**
Corn oil	8	51 ± 4	95 ± 12	97 ± 11	80 ± 9
Commercial diet	8	50 ± 5	89 ± 7	73 ± 10	75 ± 9

*$P < 0.05$.

**$P < 0.01$ (comparison by a modified Student's two-tailed t-test [Snedecor and Cochran, 1967] with the rabbits fed corn oil).

From Beynen and West]1981].

semipurified diet containing casein, a rapid increase in the concentration of serum cholesterol occurred. This elevation was mainly due to the LDL fraction and was observed within one day of feeding the semipurified casein diet (Table 3-VII). When the casein in the diet was replaced by soybean protein, a minor elevation in serum and LDL cholesterol was found. In both the casein and soybean protein-fed rabbits, low levels of VLDL cholesterol were maintained. Nevertheless, in both dietary groups the ratio of cholesterol to protein in the VLDL was markedly increased. Moreover, this ratio was higher in the rabbits fed casein than those fed soybean protein. This suggests that in rabbits fed casein, VLDL particles are synthesized with a different composition than in rabbits fed soybean protein. Since VLDL particles are assumed to be metabolized to LDL particles [Eisenberg, 1979], LDL particles with a high cholesterol:protein ratio will be formed. It is noteworthy that the most striking changes in the composition of the lipoproteins occurred during the changeover from the commercial diet to the semipurified diets. During this period a steep increase in the ratio of cholesterol to protein was found. Later, relatively minor alterations in this ratio were observed, despite changes within each dietary group in the levels of cholesterol. This might indicate that initially the essential change was in the composition of the lipoprotein particles, whereas subsequently, changes in the number of the lipoprotein particles were probably a more relevant factor.

In the above study, the serum cholesterol levels in the casein-fed rabbits reached a maximum level of about 150 mg/100 ml. In order to examine whether a further increase of the serum cholesterol levels would also result in further changes in the distribution of cholesterol between the various lipoprotein fractions, in a subsequent study rabbits were fed diets containing three levels of casein, ie, 10%, 20%, and 40% casein. The increasing proportions of dietary casein were associated with elevations in serum cholesterol levels. Table 3-VIII shows that in the animals with the highest values of serum cholesterol (the 40% casein group), most of the serum cholesterol was transported in the VLDL, whereas with moderate hypercholesterolemia (the 20% casein group) LDL was the main carrier of cholesterol.

Further, the time course of the regression and progression of hypercholesterolemia induced by semipurified diets containing 40% soybean protein and 40% casein, respectively, has also been studied. Rabbits were fed on these diets for a period of three weeks. After this period, the group receiving casein was changed to the diet containing soybean protein, whereas the animals fed the soybean protein were switched to the casein diet. The changes in the concentration of cholesterol in the various lipoprotein fractions, which occurred after the crossover, are presented in Table 3-IX. A decrease in serum cholesterol levels, due to feeding soybean protein, was reflected initially in a decrease of VLDL cholesterol, followed by a lowering in LDL cholesterol. On the other

TABLE 3-VII. Concentration of Serum and Lipoprotein Cholesterol and the Ratio of Cholesterol to Protein in the Lipoprotein Fractions in Rabbits Fed Semipurified Diets Containing Either Casein or Soybean Protein

	Initial	Day 1	Day 7	Day 31
Cholesterol[a]				
Casein				
VLDL	5 ± 2	4 ± 1	11 ± 3	12.3 ± 4
LDL	12 ± 1 *****	49 ± 5 *	71 ± 15	76.6 ± 12 *
HDL	17 ± 2 ****	35 ± 3	48 ± 6	49.9 ± 6
Serum	40 ± 4 *****	91 ± 5 *	140 ± 21	146.1 ± 15 **
Soybean protein				
VLDL	7 ± 2	4 ± 1	8 ± 2	6 ± 1
LDL	15 ± 2 ***	32 ± 4	46 ± 5	27 ± 2
HDL	20 ± 2	27 ± 4	42 ± 8	35 ± 7
Serum	46 ± 3 ****	68 ± 4	100 ± 14	72 ± 4
Cholesterol:protein				
Casein				
VLDL	0.65 ± 0.14***	1.25 ± 0.18	2.03 ± 0.23*	1.79 ± 0.31
LDL	0.54 ± 0.07******	1.47 ± 0.06	1.34 ± 0.10	1.39 ± 0.07
HDL	0.20 ± 0.03****	0.37 ± 0.02*	0.38 ± 0.05	0.33 ± 0.03
Soybean protein				
VLDL	0.61 ± 0.09***	1.01 ± 0.05	1.23 ± 0.15	1.65 ± 0.30
LDL	0.63 ± 0.07****	1.29 ± 0.10	1.26 ± 0.06	1.30 ± 0.24
HDL	0.21 ± 0.02***	0.29 ± 0.02	0.29 ± 0.02	0.31 ± 0.04

[a]Values of cholesterol expressed in mg/100 ml of whole serum and the ratio on a weight basis, mean ± SEM, six animals per group. Comparison by a modified Student's two-tailed t-test [Snedecor and Cochran, 1967] with rabbits fed diets containing soybean protein: *$P < 0.05$; **$P < 0.01$; with rabbits on the same diet at Day 1: ***$P < 0.05$; ****$P < 0.01$; *****$P < 0.001$. From Terpstra and Sanchez-Muniz [1981].

TABLE 3-VIII. Concentration of Cholesterol in Serum Lipoproteins From Rabbits Fed Semipurified Diets Containing Different Proportions of Casein†

| | Cholesterol concentration (mg/100 ml, mean ± SEM) | | | |
| | | Semipurified diets | | |
	Initial	10% casein	20% casein	40% casein
VLDL	18 ± 3	41 ± 5*	119 ± 34	542 ± 197*
LDL	14 ± 2	31 ± 14**	147 ± 15	210 ± 57
HDL	18 ± 1	15 ± 3***	60 ± 6	45 ± 12
Serum	59 ± 4	112 ± 19**	372 ± 53	958 ± 282

†The diets were fed for 28 days; each group comprised six animals; the initial values represent the average values of all the 18 rabbits immediately before the change to the semipurified diets. Statistical comparison by a modified two-tailed t-test [Snedecor and Cochran, 1967] with the 20% casein group.
*$P < 0.05$.
**$P < 0.01$.
***$P < 0.001$.
From Terpstra et al [1981a].

TABLE 3-IX. Concentration of Cholesterol in Serum Lipoprotein Fractions From Rabbits Fed Semipurified Diets Containing Either Casein or Soybean Protein†

	Day 0	Day 1	Day 2	Day 4	Day 7	Day 10
Soybean protein						
VLDL	94 ± 24	78 ± 32	45 ± 16	35 ± 22	23 ± 5	18 ± 5
LDL	167 ± 45	167 ± 44	154 ± 41	126 ± 41	93 ± 41	62 ± 25
HDL	26 ± 4	35 ± 4	39 ± 2	48 ± 6	39 ± 5	29 ± 5
Serum	309 ± 59	284 ± 63	254 ± 62	204 ± 59	153 ± 41	111 ± 28
Casein						
VLDL	11 ± 2**	11 ± 1*	12 ± 2*	18 ± 3	19 ± 5	21 ± 5
LDL	27 ± 9*	46 ± 10*	52 ± 10*	66 ± 15	70 ± 23	87 ± 30
HDL	37 ± 7	40 ± 8	45 ± 7	46 ± 6	51 ± 9	52 ± 10
Serum	80 ± 14*	97 ± 15*	111 ± 16	120 ± 20	136 ± 24	165 ± 41

†Values expressed in mg/100 ml of whole serum, mean ± SEM, six rabbits per group. The rabbits on the casein diet had been fed previously a diet containing soybean protein for a period of three weeks and were switched to the casein diet on day 0. Conversely, the rabbits on the soybean protein diet had been fed a casein diet and were changed to the soybean protein diet on day 0. Statistical comparison by a modified Student's two-tailed test [Snedecor and Cochran, 1967] with the soybean protein group.
*$P < 0.05$.
**$P < 0.01$.
From Terpstra et al [1982b].

hand, the progressive elevation in serum cholesterol in the rabbits switched from the soybean protein diet to the casein diet was mainly caused by an increase in LDL cholesterol. Thus, these studies show that the induction of hypercholesterolemia in rabbits fed casein is initially reflected in an elevated level of LDL cholesterol followed by a subsequent increase in VLDL.

In a further study, a comparison was made between hypercholesterolemia induced by feeding semipurified diets containing soybean protein plus cholesterol or casein. It was found that cholesterol- and casein-induced hypercholesterolemia develop in a similar way: At similar levels of total serum cholesterol, a similar distribution of cholesterol between the VLDL and LDL fractions was observed (Fig. 3-X). In addition, the rabbits fed cholesterol-enriched soybean protein diets and cholesterol-free casein diets showed a marked increase of apo E in the VLDL fraction (Fig. 3-XI).

It is interesting to note the large changes in the density profile of the serum lipoproteins which occur in rabbits fed semipurified diets. When rabbits were fed semipurified diets containing either casein or soybean protein, marked

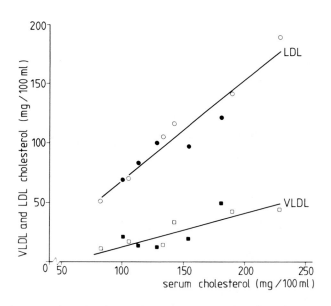

Fig. 3-X.   Concentration of cholesterol in VLDL, LDL, and whole serum in rabbits fed semipurified diets containing either soybean protein plus cholesterol or casein: LDL (○) and VLDL (□) cholesterol in cholesterol-fed rabbits; LDL (●) and VLDL (■) in casein-fed animals. The regression lines were calculated according to the following formula: VLDL (or LDL) cholesterol = (total serum cholesterol) a + b (from Scholz et al [1982] and unpublished observations by the same authors).

variations in the density of the LDL and HDL were found between individual rabbits (Fig. 3-XII). No such variation between rabbits was observed when commercial diets were fed. Thus, the density limits defined for human serum lipoproteins are not equally applicable to the lipoproteins of hypercholesterolemic rabbits. It might be more appropriate to collect the lipoproteins of both normo- and hypercholesterolemic rabbits on the basis of their appearance in the gradient [Terpstra and Sanchez-Muniz, 1981] or to collect fractions with narrower density limits [Scholz et al, 1982] rather than to use the conventional density limits as defined for human serum lipoproteins.

These findings also throw some light on the different results found by several authors on the relative concentration of cholesterol in the intermediate-density lipoprotein (IDL) and LDL fractions, as defined for human serum lipoproteins, in rabbits fed casein. Huff et al [1982] reported that the increase of cholesterol in the density (d) range of $1.006 < d < 1.063$ was mainly attributable to increased levels of IDL ($1.006 \, d < 1.019$). On the other hand, Ross et al [1978] and Lacombe and Nibbelink [1980] reported higher levels of cholesterol in the LDL ($1.019 < d < 1.063$) compared with the IDL fractions ($1.006 < d < 1.019$). However, these contradictory results might be explained by the large variations between individual rabbits in the density profile of the

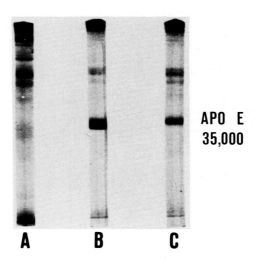

Fig. 3-XI.   Sodium dodecyl sulfate-gel electrophoresis patterns of VLDL apoproteins in 10% polyacrylamide gels. Lipoproteins were isolated from the serum of rabbits fed semipurified diets containing soybean protein (A), soybean protein plus cholesterol (B), or casein (C). Blood samples were taken after feeding the diets for a period of four weeks (from Scholz et al [1982]).

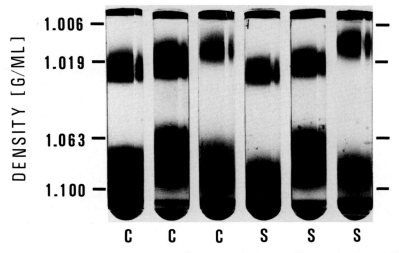

Fig. 3-XII. Photograph of the density profile of prestained serum lipoproteins from rabbits fed semipurified diets containing either casein (C) or soybean protein (S). This picture clearly shows the differences between individual rabbits in the density profile of the serum lipoproteins when fed semipurified diets (from Terpstra and Sanchez-Muniz [1981]).

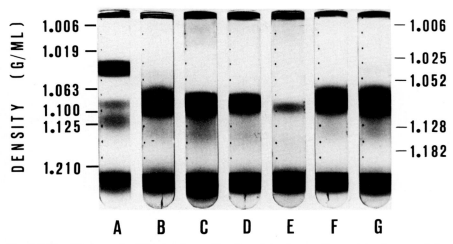

Fig. 3-XIII. Photograph of the density profile of prestained serum lipoproteins from pooled sera of lean female Zucker rats fed various diets compared to that of a normocholesterolemic human. A) Normocholesterolemic human. B) Rats fed a commercial diet. C) Rats fed a cholesterol-enriched commercial diet. Rats fed a cholesterol-enriched semipurified diet containing 20% casein (D), 50% casein (E), 20% soybean protein (F), and 50% soybean protein (G). The semipurified diets and the cholesterol-enriched commercial diet contained 1.2% cholesterol (from Terpstra et al [1982e]).

lipoproteins and perhaps also by differences in the distribution of lipoproteins in blood samples obtained at various times after feeding.

B. Rats

Rats have a serum lipoprotein pattern which is quite different from that of man and rabbits. Figures 3-XIII and 3-XIV show the density profile of the prestained serum lipoproteins of normocholesterolemic lean and genetically obese Zucker rats compared with those of a normocholesterolemic human. Lean Zucker rats exhibit a broad lipoprotein band in the density range of the HDL as defined for human serum lipoproteins (Fig. 3-XIII). No LDL band can be discerned. Genetically obese Zucker rats, however, have a clearly visible LDL and HDL band (Fig. 3-XIV).

When lean female Zucker rats were fed a cholesterol-enriched semipurified diet containing casein, a marked elevation in serum cholesterol level occurred. Low levels of serum cholesterol were maintained when soybean protein was included in the diet. These changes in serum cholesterol were associated with marked alterations in the density profile of the serum lipoproteins and the distribution of cholesterol between the various lipoprotein classes (Fig. 3-XIII). The rats fed casein exhibited an HDL band which was much less pronounced than in the rats fed soybean protein. Moreover, this band was even fainter

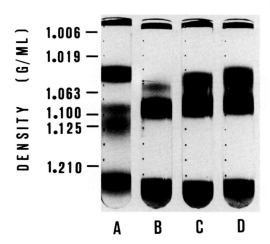

Fig. 3-XIV. Photograph of the density profile of the prestained serum lipoproteins from pooled sera of genetically obese female Zucker rats fed various diets compared to that of a normocholesterolemic human serum sample. A) Normocholesterolemic human. B) Rats fed a commercial diet. Rats fed a cholesterol-enriched semipurified diet containing soybean protein (C) and casein (D). The semipurified diets contained 50% protein and 1.2% cholesterol (from Terpstra et al [1983]).

when a 50% casein diet was fed instead of a 20% casein diet. The visual impressions were reflected in the cholesterol concentrations of the various lipoprotein fractions (Table 3-X). When dietary soybean protein was replaced by casein, there was an increased amount of cholesterol in the lipoproteins with a low density, particularly the VLDL fraction. This was associated with a lowering of the cholesterol level in the HDL fraction. A more or less similar pattern of response has been reported in man. Van Raaij et al [1981, 1982] found a significant increase in HDL cholesterol and a decrease in LDL cholesterol when soybean protein replaced casein in the diets of normocholesterolemic subjects.

Quite a different pattern of response to dietary casein and soybean protein was found in genetically obese Zucker rats when compared with their lean counterparts. The feeding of cholesterol-enriched semipurified diets resulted in the appearance of a pronounced LDL band (Fig. 3-XIV). The distribution of cholesterol between the various lipoprotein fractions is presented in Table 3-XI. In the rats fed a commercial diet, most of the serum cholesterol is transported in the HDL fraction. When the cholesterol-enriched semipurified diet containing soybean protein was fed, an increase in both the HDL and LDL fraction occurred. However, when a further increase in total serum cholesterol was induced by feeding the casein diet, a relatively small elevation in HDL cholesterol was seen together with a marked increase in LDL and VLDL cholesterol.

Thus, all these studies in rats and also those in rabbits show that the nature and the type of dietary protein can profoundly affect the density profile of and the cholesterol concentration in the serum lipoproteins.

TABLE 3-X. Concentration of Cholesterol in Serum Lipoprotein Fractions From Lean Female Zucker Rats Fed Cholesterol-Enriched, Semipurified Diets Containing Different Proportions of Either Casein or Soybean Protein*

Lipoprotein fraction	Soybean protein		Casein	
	50%	20%	20%	50%
d < 1.006	3	5	62	599
1.006 < d < 1.025	2	2	9	24
1.025 < d < 1.052	3	3	6	9
1.052 < d < 1.128	42	48	31	21
1.128 < d < 1.182	6	6	6	4
1.182 < d	4	3	3	4
Serum	63	75	165	737

*The concentration of cholesterol is expressed in mg/100 ml of whole serum. The lipoproteins were separated by density-gradient ultracentrifugation from pools of eight animals in each group, after feeding the diets for a period of 14 weeks. The densities of the lipoprotein fractions are expressed in g/ml.
From Terpstra et al [1982c].

IV. DIETARY PROTEIN AND ATHEROSCLEROSIS

High levels of serum cholesterol are considered a risk factor in the genesis of atherosclerosis [Stamler, 1979]. Therefore the degree of atherosclerosis was studied in rabbits fed casein and various amino acids. Figure 3-XV shows that modulations of serum cholesterol levels due to modification of the amino acid composition of the diets was found to cause parallel changes in the proportion of aortic surface covered by atheromatous plaques. Thus, these results show that modulations of the protein source in the diet also have an impact on the etiology of atherosclerosis.

V. POSSIBLE MECHANISMS

The mechanism underlying the cholesterolemic properties of dietary protein is not yet clearly understood. However, several studies in rats [Nagata et al, 1981] and rabbits [Huff and Carroll, 1980b] have shown that the hypocholesterolemic effects of plant proteins compared to casein in the diet is mediated by an increased drainage of bile acids and neutral steroids in the feces. Further, replacement of dietary soybean protein by casein has been found to result in a lower rate of cholesterol synthesis [Reiser et al, 1977] and turnover [Huff and Carroll, 1980b].

TABLE 3-XI. Concentration of Cholesterol in Serum, Lipoproteins, and Liver in Genetically Obese Female Zucker Rats Fed Cholesterol-Enriched Semipurified Diets Containing Either Casein or Soybean Protein†

		Semipurified diet	
	Commercial diet	Soy protein	Casein
Serum cholesterol (mg/100 ml serum)			
Initial	100 ± 3	101 ± 6	103 ± 3
5 weeks	79 ± 4	120 ± 10	409 ± 51*
11 weeks	86 ± 4	145 ± 10	325 ± 23**
Lipoprotein cholesterol (mg/100 ml serum)			
VLDL d < 1.006	19	12	87
LDL 1.006 < d < 1.063	13	38	75
HDL 1.063 < d < 1.21	51	98	112
Liver cholesterol (g/100 g wet weight)	0.20 ± 0.01	0.58 ± 0.05	6.74 ± 0.60**

†Values are expressed as mean ± SEM, five animals per group. The semipurified diets contained 50% protein and 1.2% cholesterol. Lipoprotein and liver cholesterol were determined after feeding the diets for 11 weeks: Lipoproteins were isolated from sera pooled per group. Statistical comparison by a modified Student's two-tailed test [Snedecor and Cochran, 1967] with the soybean protein group.
*$P < 0.01$.
**$P < 0.001$.
From Terpstra et al [1983].

The major pathway for the disposal of cholesterol from the body is the excretion of bile acids and neutral steroids in the feces [Dietschy, 1968]. The secretion and reabsorption of the bile acids and cholesterol by means of the enterohepatic circulation plays an important role in the regulation of cholesterol metabolism. Bile acids are produced from cholesterol in the liver and are subsequently excreted in the bile. However, most of the bile acid is reabsorbed from the intestine and recycled. Thus, a less efficient reabsorption of bile acids leads to an increased excretion in the feces, resulting in an enhanced removal of cholesterol from the body. Further, there is evidence that changes in the excretion of bile acids also affects the serum cholesterol levels. In our laboratory, it has been found that in casein-fed rabbits the feeding of cholestyramine, a bile acid sequestrant, effectively reduced the levels in serum of cholesterol and also of triglycerides (Table 3-XII). Similarly in hypercholesterolemic patients, ileal bypass has been reported to result in an enhanced excretion of bile acids together with a decrease of serum cholesterol levels [Buchwald et al, 1974]. On the other hand, rabbits starved for several days, in which fecal production and hence the fecal excretion of bile acids and neutral steroids are markedly reduced, are known to develop severe hypercholesterolemia [Klauda and Zilversmit, 1974; Swanor and Connor, 1975].

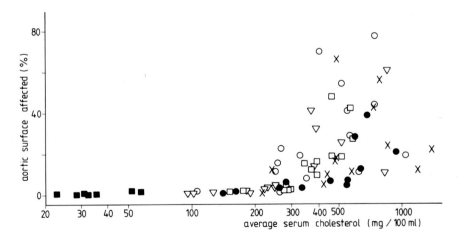

Fig. 3-XV.   Proportion of arotic surface covered by plaques as a function of serum cholesterol concentration in rabbits fed for 49 weeks on semipurified diets containing different proteins and supplements of various amino acids. ×, 20.8% casein; ●, 20.8% casein + 0.35% arginine + 0.54% alanine; ∇, 20.8% casein + 0.35% arginine + 0.54% alanine + 0.65% glycine; ○, 20.8% casein + 0.35% arginine + 0.54% alanine + 1.40% glycine; □, 20.8% casein + 0.35% arginine + 0.54% alanine + 3.15% glycine; ■, commercial diet (from Katan et al [1982]).

However, it is not clear how dietary protein can affect the excretion of bile acids. It is possible that the rate of digestion of the proteins plays an important role. Roy and Schneeman [1981] observed that mice fed soybean protein had higher intestinal contents than their counterparts fed casein, suggesting a slower digestion of soybean protein compared with casein. Further, it has been reported that proteins which are not completely digested interfere with the re-absorption of bile acids from the intestine [Sklan et al, 1979; Sklan, 1980]. This might mean that the feeding of soybean protein results in a less efficient reabsorption of bile acids from the intestine. In particular, the presence of un-digested soybean protein in the distal part of the intestine can be crucial, since most of the bile acids are reabsorbed in this region [Dietschy and Wilson, 1970]. Interference of the reabsorption of bile acids by dietary soybean protein may lead to an enhanced excretion of bile acids in the feces. This may increase cholesterol catabolism in the liver [Dietschy and Wilson, 1970]. The data from Nagata et al [1981] are in line with this hypothesis. They reported that rats fed casein excreted less fecal bile acids than rats on a soybean protein diet. How-ever when mixtures of amino acids with an amino acid composition similar to casein or soybean protein were fed, these differences in bile acid excretion dis-appeared. Further, Huff and Carroll [1980a] reported that an amino acid mix-ture equivalent in amino acid composition to that of soybean protein produced concentrations of cholesterol in the serum higher than those obtained with in-tact soybean protein. The addition of amino acids to casein and soybean pro-tein in order to provide an amino acid composition equivalent to soybean pro-tein and casein, respectively, supported the hypothesis that the intact protein component of each mixture had an overriding effect.

TABLE 3-XII. Effect of Cholestyramine on the Concentration of Cholesterol and Triglycerides in the Serum of Casein-Fed Rabbits†

	Concentration (mg/100 ml, mean ± SEM)	
Diet	Cholesterol	Triglycerides
Casein	572 ± 72	497 ± 90
Casein plus 2% cholestyramine		
1 week	221 ± 42**	115 ± 24**
2 weeks	93 ± 11**	112 ± 23**
3 weeks	102 ± 7**	117 ± 46*

†The casein diet was fed for a period of six months to a group of six rabbits which were then transferred to a casein diet containing 2% cholestyramine. Statistical comparison by a modified Student's two-tailed t-test [Snedecor and Cochran, 1967] with the casein diet without choles-tyramine.
*$P < 0.05$
**$P < 0.01$.
From Dallinga-Thie G, De Rooij C, Hermus RJJ, Katan MB [unpublished observations].

Nevertheless, several studies have shown that supplementation of dietary proteins with some particular amino acids can also profoundly affect the levels of serum cholesterol. In our laboratory, we have observed that in both rabbits (Fig. 3-IV) and rats (Fig. 3-V) the addition of glycine to a casein diet resulted in a decrease in serum cholesterol levels compared with unsupplemented protein. This might indicate that some amino acids or a combination of amino acids are able to influence cholesterol metabolism. However, other studies, in which mixtures of amino acids alone or proteins supplemented with amino acids were fed, did not provide clear-cut results [Huff and Carroll, 1980a].

VI. SUMMARY

In rabbits, marked changes in the levels of serum cholesterol could be induced by feeding cholesterol-free diets containing different proteins. The cholesterolemic effect of dietary proteins was enhanced by increasing the proportion of protein in the diet. Changes in the levels of serum cholesterol could be achieved by modification in the amino acid composition of the diets. In particular, the addition of glycine to a casein diet resulted in a marked reduction in the levels of serum cholesterol. Elevations in serum cholesterol were initially reflected in an increased level of LDL cholesterol, but at higher concentrations of serum cholesterol the VLDL was the main carrier of cholesterol. The feeding of semipurified diets to rabbits resulted in marked variations in the density profile of the serum lipoproteins between individual rabbits. Furthermore, the cholesterolemic response to dietary protein is largely reduced in mature adult rabbits compared to their young growing counterparts. Finally, the type of fat and fiber in the diet can greatly affect the results.

In rats, results similar to those observed in rabbits were obtained when cholesterol-enriched diets were used. However, in genetically obese Zucker rats, an elevation of serum cholesterol due to feeding casein was reflected in a cholesterol elevation in all the lipoprotein fractions, whereas in the lean rats the changes in serum cholesterol were mainly attributable to alterations in VLDL cholesterol. Further, female rats were found to be more susceptible than males to changes in serum cholesterol levels produced by feeding various proteins.

Thus, in rabbits and rats the feeding of various dietary proteins can profoundly affect the levels of serum cholesterol. However, the mechanism underlying this phenomenon has not been established; therefore, it is not clear whether these findings can be extrapolated to man.

VII. REFERENCES

Anitschkow N, Chalatow S (1913): Ueber experimentelle Cholesterin-steatose und ihre Bedeutung für die Entstehung einiger pathologischer Prozesse. Zentralbl Allg Pathol Pathology Anat 24:1–9.

Beynen AC, West CE (1981): The distribution of cholesterol between lipoprotein fractions of serum from rabbits fed semipurified diets containing casein and either coconut oil or corn oil. Z Tierphysiol Tierernaehrg Futtermittelkd 46:233–239.

Beynen AC, van Wanrooij-Stroeken CTM (1981): Relations between dietary salt type, acidosis and hyperlipemia in rabbits on casein containing semipurified diets. Z Tierphysiol Tierernaehrg Futtermittelkd 46:240–246.

Buchwald H, Moore HB, Varco RL (1974): Ten years clinical experience with partial ileal bypass in management of the hyperlipidemias. Ann Surg 180:384–392.

Carroll KK, Hamilton RMG (1975): Effects of dietary protein and carbohydrate on plasma cholesterol levels in relation to atherosclerosis. J Food Sci 40:18–23.

Carroll KK, Giovannetti PM, Huff MW, Moase O, Roberts DCK, Wolfe BM (1978): Hypocholesterolemic effect of substituting soybean protein for animal protein in the diet of healthy young women. Am J Clin Nutr 31:1312–1321.

Connor WE, Connor SL (1972): The key role of nutritional factors in the prevention of coronary heart disease. Prev Med 1:49–83.

Descovich GC, Gaddi A, Mannino G, Cattin L, Senin U, Caruzzo C, Fragiocomo C, Sirtori M, Ceredi C, Benassi MS, Colombo L, Fontana G, Mannarino E, Bertelli E, Noseda G, Sirtori CR (1980): Multicentre study of soybean protein diet for outpatient hypercholesterolaemic patients. Lancet ii:709–712.

Dietschy JM (1968): Mechanism for the intestinal absorption of bile acids. J Lipid Res 9:297–309.

Dietschy JM, Wilson JD (1970): Regulation of cholesterol metabolism. Third of three parts. N Engl J Med 282:1241–1249.

Eisenberg S (1979): Very-low-density lipoprotein metabolism. In Eisenberg S (ed): "Lipoprotein Metabolism." Basel: S. Karger, pp 139–165.

Grundy SM (1979): Dietary fats and sterols. In Levy RI, Rifkind BM, Dennis BH, Ernst ND (eds): "Nutrition, Lipids and Coronary Heart Disease." New York: Raven Press, pp 89–118.

Guigoz Y, Stasse-Wolthuis M, Hermus RJJ (1979): Sources de protéines, taux de cholestérol et composition des lipoprotéins du rat, normal et obese. Int J Vitam Nutr Res 20:32–42.

Hermus RJJ (1975): "Experimental Atherosclerosis in Rabbits on Diets With Milk Fat and Different Proteins." Wageningen, The Netherlands: Centre for Agricultural Publishing and Documentation.

Hermus RJJ, Stasse-Wolthuis M (1978): Lipids and lipoproteins in rabbits fed semisynthetic diets containing different proteins. In Peeters H (ed): "Protides of the Biological Fluids." Oxford: Pergamon Press, pp 457–460.

Hermus RJJ, Terpstra AHM, Dallinga-Thie GM (1979): Aanwijzingen voor een rol van voedingseiwitten bij de beinvloeding van het serum-cholesterolgehalte. Voeding 40:95–99.

Hermus RJJ, Dallinga-Thie GM (1979): Soya, saponins, and plasma cholesterol. Lancet i:48.

Huff MW, Hamilton RMG, Carroll KK (1979): Plasma cholesterol levels in rabbits fed a low fat, cholesterol-free semipurified diet: Effects of dietary proteins, protein hydrolysates and amino acid mixtures. Atherosclerosis 28:187–195.

Huff MW, Carroll KK (1980a): Effects of dietary proteins and amino acid mixtures on plasma cholesterol levels in rabbits. J Nutr 110:1676–1685.

Huff MW, Carroll KK (1980b): Effects of dietary protein on turnover, oxidation, and absorption of cholesterol, and on steroid excretion in rabbits. J Lipid Res 21:546–558.

Huff MW, Roberts DCK, Carroll KK (1982): Long term effects of semipurified diets containing casein or soy protein isolate on atherosclerosis and plasma lipoproteins in rabbits. Atherosclerosis 41:327–336.

Ignatowski A (1909): Ueber die Wirkung des tierischen Eiweisses auf die Aorte und die parenchymatösen Organe der Kaninchen. Virchows Arch Pathol Anat Physiol Klin Med 198:248–270.

Katan MB, Vroomen L, Hermus RJJ (1982): Reduction of casein induced hypercholesterolemia and atherosclerosis in rabbits and rats by dietary glycine. Atherosclerosis 43:381–391.

Kim DN, Lee KT, Reiner JM, Thomas WA (1978): Effects of a soy protein product on serum and tissue cholesterol concentrations in swine fed high-fat, high-cholesterol diets. Exp Mol Pathol 29:385–399.

Klauda HC, Zilversmit DB (1974): Influx of cholesterol into plasma in rabbits with fasting hyperbetalipoproteinemia. J Lipid Res 15:593–601.

Kritchevsky D (1964): Experimental atherosclerosis in rabbits fed cholesterol-free diets. J Atheroscler Res 4:103–105.

Kritchevsky D, Tepper SA, Williams DE, Story JA (1977): Experimental atherosclerosis in rabbits fed cholesterol-free diets. Part 7. Interaction of animal or vegetable protein with fiber. Atherosclerosis 26:397–403.

Kritchevsky D (1979): Vegetable protein and atherosclerosis. J Am Oil Chem Soc 56:135–140.

Lacombe C, Nibbelink M (1980): Lipoprotein modifications with changing dietary proteins in rabbits on a high fat diet. Artery 6:280–289.

Meeker DR, Kesten MD (1941): Effect of high protein diets on experimental atherosclerosis in rabbits. Arch Pathol 31:147–162.

Moyer AW, Kritchevsky D, Logan JB, Cox HR (1956): Dietary protein and serum cholesterol in rats. Proc Soc Exp Biol Med 92:736–737.

Nagata Y, Tanaka K, Sugano M (1981): Further studies on the hypocholesterolaemic effect of soya-bean protein in rats. Br J Nutr 45:233–241.

Nath N, Harper AE, Elvehjem CA (1959): Diet and cholesteremia. Part 3. Effect of dietary proteins with particular reference to the lipids in wheat gluten. Can J Biochem Physiol 37:1375–1384.

Newburgh LH (1919): The production of Bright's disease by feeding high protein diets. Arch Intern Med 24:359–377.

Raaij JMA van, Katan MB, Hautvast JGAJ, Hermus RJJ (1981): Effects of casein versus soy protein diets on serum cholesterol and lipoproteins in young healthy volunteers. Am J Clin Nutr 34:1261–1271.

Raaij JMA van, Katan MB, West CE, Hautvast JGAJ (1982): Influence of diets containing casein, soy isolate and soy concentrate on serum cholesterol and lipoproteins in middle-aged volunteers. Am J Clin Nutr 35:925–934.

Reiser R, Henderson GR, O'Brien BC, Thomas J (1977): Hepatic 3-hydroxy-3-methylglutaryl coenzyme-A reductase of rats fed semipurified and stock diets. J Nutr 107:453–457.

Ross AC, Minick CR, Zilversmit DB (1978): Equal atherosclerosis in rabbits fed cholesterol-free, low-fat diet or cholesterol-supplemented diet. Atherosclerosis 29:301–315.

Roy DM, Schneeman BO (1981): Effect of soy protein, casein and trypsin inhibitor on cholesterol, bile acids and pancreatic enzymes in mice. J Nutr 111:878–885.

Scholz KE, Beynen AC, West CE (1982): Comparison between the hypercholesterolaemia in rabbits induced by semipurified diets containing either cholesterol or casein. Atherosclerosis 44:85–97.

Sirtori CR, Gatti E, Montero O, Conti F, Agradi E, Tremoli E, Sirtori M, Fraterrigo L, Tavazzi L, Kritchevsky D (1979): Clinical experience with the soybean protein diet in the treatment of hypercholesterolemia. Am J Clin Nutr 32:1645–1658.

Sklan D, Budowski P, Hurwitz S (1979): Absorption of oleic and taurocholic acids from the intestine of the chick. Interactions and interference by proteins. Biochim Biophys Acta 573:31–39.

Sklan D (1980): Digestion and absorption of casein at different dietary levels in the chick: Effect on fatty acid and bile acid absorption. J Nutr 110:989–994.

Snedecor GW, Cochran WG (1967): "Statistical Methods." Ames: Iowa State University Press, chap 4, pp 114–116, 130–131.

Stamler J (1979): Population studies. In Levy RI, Rifkind BM, Dennis BH, Ernst ND (eds): "Nutrition, Lipids and Coronary Heart Disease." New York: Raven Press, pp 25–88.

Stasse-Wolthuis M (1980): Influence of dietary fiber on cholesterol metabolism and colonic function in healthy subjects. World Rev Nutr Diet 36:100–140.

Swanor JC, Connor WE (1975): Hypercholesterolemia of total starvation. Its mechanism via tissue mobilization of cholesterol. Am J Physiol 229:365–369.

Terpstra AHM (1981): The effect of semipurified diets containing either casein or soybean protein on the concentration of serum cholesterol and the lipoprotein composition in rabbits. Wageningen, The Netherlands: Ph.D. Thesis.

Terpstra AHM, Sanchez-Muniz FJ (1981): Time course of the development of hypercholesterolemia in rabbits fed semipurified diets containing casein or soybean protein. Atherosclerosis 39:217–227.

Terpstra AHM, Harkes L, van der Veen FH (1981a): The effect of different proportions of casein in semi-purified diets on the concentration of serum cholesterol and the lipoprotein composition in rabbits. Lipids 16:114–119.

Terpstra AHM, Woodward CJH, Sanchez-Muniz FJ (1981b): Improved techniques for the separation of serum lipoproteins by density gradient ultracentrifugation: Visualization by prestaining and rapid separation of serum lipoproteins from small volumes of serum. Anal Biochem 111:149–157.

Terpstra AHM, Hermus RJJ, West CE (1982a): The role of dietary protein in cholesterol metabolism. World Rev Nutr Diet (in press).

Terpstra AHM, Woodward CJH, West CE, van Boven HG (1982b): A longitudinal cross-over study of serum cholesterol and lipoproteins in rabbits fed on semi-purified diets containing either casein or soya-bean protein. Br J Nutr 47:213–221.

Terpstra AHM, van Tintelen G, West CE (1982c): The effect of semipurified diets containing different proportions of either casein or soybean protein on the concentration of cholesterol in whole serum, serum lipoproteins and liver in male and female rats. Atherosclerosis 42: 85–95.

Terpstra AHM, van Tintelen G, West CE (1982d): The hypocholesterolemic effect of dietary soy protein in rats. J Nutr 112:810–817.

Terpstra AHM, Sanchez-Muniz FJ, West CE, Woodward CJH (1982e): The density profile and cholesterol concentration of serum lipoproteins in domestic and laboratory animals. Comp Biochem Physiol 71B:669–673.

Terpstra AHM, van Tintelen G, West CE (1983): Serum lipids, lipoprotein composition and liver cholesterol in genetically obese Zucker rats fed semipurified diets containing either casein or soy protein. Ann Nutr Metab (in press).

Wacker L, Hueck W (1913): Ueber experimentelle Atherosclerose und Cholesterinaemie. Muenchener Med Wochenschr 60:2097–2100.

West CE, Deuring K, Schutte JB, Terpstra AHM (1982): The effect of age on the development of hypercholesterolemia in rabbits fed semipurified diets containing casein. J Nutr 112:1287–1295.

Wolfe BM, Giovannetti PM, Cheng DCH, Roberts DCK, Carroll KK (1981): Hypolipidemic effect of substituting soybean protein isolate for all meat and dairy protein in the diets of hypocholesterolemic men. Nutr Rep Int 24:1187–1198.

Yerushalmy J, Hilleboe HE (1957): Fat in the diet and mortality from heart disease. A methodologic note. NY State J Med 57:2343–2534.

Zilversmit DB (1979): Dietary fiber. In Levy RI, Rifkind BM, Dennis BH, Ernst ND (eds): "Nutrition, Lipids and Coronary Heart Disease." New York: Raven Press, pp 149–174.

*Animal and Vegetable Proteins in Lipid
Metabolism and Atherosclerosis, pages 51–84*
© *1983 Alan R. Liss, Inc., 150 Fifth Ave., New York, NY 10011*

4
Hypocholesterolemic Effect of Plant Protein in Relation to Animal Protein: Mechanism of Action

Michihiro Sugano

Laboratory of Nutrition Chemistry, Kyushu University School of Agriculture, Fukuoka 812, Japan

I. INTRODUCTION

Though the concept of the protein effect emerged more than 70 years ago [Ignatowski, 1909], protein was probably the last nutrient designated as a dietary element that regulates plasma lipid levels. Epidemiological surveys have also pointed to dietary protein as a potent determinant in controlling the level of plasma cholesterol and hence mortality from coronary heart diseases [Yudkin, 1957; Connor and Connor, 1972]. Recent accumulating evidence generally favors the concept that proteins of plant origin produce lower plasma levels of cholesterol and possibly triglycerides in human as well as animal models [Carroll and Hamilton, 1975; Kritchevsky, 1979; Carroll, 1981; Terpstra et al, 1982]. Regarding the dietary treatment of hypercholesterolemia, the substitution of plant proteins for their animal counterparts seems compatible with normal eating habits, since current food technology has enough potential to provide various forms of plant protein preparations, especially from soybean protein, intended for human consumption as animal protein substitutes [Waggle and Kolar, 1979].

Despite considerable amounts of knowledge concerning the cholesterol-lowering efficacy of plant proteins, there is still no conclusive agreement concerning the mechanism involved in this action. This is especially true for humans. In this chapter, the author focuses on the mechanism of the hypocholesterolemic efficacy of plant proteins on the basis of available information obtained almost exclusively from animal models. This includes effects of plant proteins, in most cases soy protein, on several parameters of cholesterol metabolism. Effects of dietary proteins on apolipoprotein metabolism will also be briefly discussed. In addition, effects of differences in amino acid composition between plant and animal proteins are examined in the light of the cholesterol-lowering action.

Since hypercholesterolemia can result from a disproportion of cholesterol balance in the body, insight into parameters of cholesterol metabolism will provide us with an important clue to the mechanism responsible for provoking the cholesterol-lowering action of plant proteins. Available evidence is indicative of the diversity of response; not only species-specific differences, but also the marked influence of experimental conditions such as the nature of protein preparations or compositions of diets employed must be considered. Thus, the reported data must be evaluated with the greatest care.

For phenomenalistic observations, the readers should refer to current impressive reviews [Carroll and Hamilton, 1975; Kritchevsky, 1979; Carroll, 1981; Terpstra et al, 1982]. Owing to the species specificity of cholesterol metabolism, the present paper does not cover all of the available data, but is restricted to the observations obtained in humans, rabbits, and rats under reasonable dietary protein levels. Studies on the protein effect in other animal species as well as those with low-protein diets are beyond the scope of the present discussion.

II. EFFECTS ON ABSORPTION AND EXCRETION OF CHOLESTEROL AND BILE ACIDS

Fumagalli et al [1978] are perhaps the first investigators who provided comparable data showing the effect of dietary protein on the fecal output of neutral steroids. They fed cholesterol-free, semipurified diets with either casein or soybean meal as the source of dietary protein to rabbits and observed an increased fecal excretion of neutral steroids but not acidic steroids after soybean meal. Although the Italian investigators considered this observation in connection with the hypocholesterolemic action of soy protein, it was diffi-

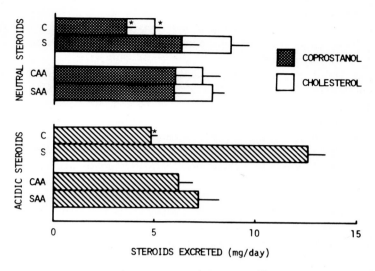

Fig. 4-I.   Effect of dietary proteins and amino acid mixtures equivalent to proteins on fecal excretion of neutral and acidic steroids in rats fed low-fat, cholesterol-free, semipurified diets [Nagata et al, 1981a]. C, casein; S, soy protein; CAA, casein amino acid mixture; SAA, soy protein amino acid mixture. *Significantly different from the corresponding soy protein group.

cult to link the increased excretion to either the protein component itself or soy polysaccharide; their soybean meal contained the protein only at the 50–60% level and polysaccharide at approximately 20%.

This uncertainty was soon resolved by using more purified soybean protein preparations. Sautier et al [1979] demonstrated in rats fed cholesterol-free, semipurified diets containing casein or the soy protein isolate (containing more than 90% protein on a dry-matter basis; this level is comparable with that of casein) that the dry weight of the feces and daily excretion of neutral steroids were greater with the soy protein diet than with the casein diet. Soy protein isolate also measurably, but not significantly, increased bile acid excretion. Strangely, the plasma cholesterol level remained unchanged even under increased fecal excretion of steroids. Nagata et al [1980] confirmed the increased neutral steroid excretion in more systemic studies with rats fed a low-fat, cholesterol-free, semipurified diet; and, as shown in Figure 4-I, they subsequently showed a significant increase in acidic as well as neutral steroids on feeding the soy protein isolate relative to casein [Nagata et al, 1981a]. The observation that soy protein substituted for casein in the lithogenic diet repressed gallstone formation and promoted dissolution of preestablished gallstones in hamsters [Kritchevsky and Klurfeld, 1979] appears to be the result of increased fecal steroid excretion.

Whether soybean protein likewise raises fecal output of steroids in these rodents fed cholesterol-enriched diets is uncertain. In swine, Kim et al [1978] were initially unable to show any differences in the fecal excretion of neutral and acidic steroid when the effect of casein and soybean protein preparation (approximately 67% protein on a dry-matter basis) was compared in semipurified hyperlipidemic diets high in fat and cholesterol. The mean value of cholesterol absorption for the soy protein group was rather higher than for the casein group, though not significant. In subsequent studies using a more purified soy protein preparation (92% purity) in a similar type of semipurified hyperlipidemic diet, Kim et al [1980a] showed a significant increase in the fecal excretion of both neutral and acidic steroids after switching from casein to soy protein.

In contrast to the increase in fecal output of bile acids as well as neutral steroids in rats fed soy protein compared with those fed casein [Nagata et al, 1981a], these differences disappeared when the corresponding amino acid mixtures were fed instead of the intact proteins, though the serum level of cholesterol was consistently, but to a somewhat lesser extent as compared with the intact proteins, lower on feeding the soy protein-type amino acid mixture than the casein-type counterpart (Fig. 4-I). The significance of this observation will be discussed later with regard to the mechanism of the serum cholesterol-lowering action of soy protein in humans, but the results suggest that the soy protein effect is not totally explained by increased steroid excretion alone.

Using a sophisticated approach, more dynamic aspects of the parameters of cholesterol balance were recently studied in rabbits by Huff and Carroll [1980a]. Rabbits fed a low-fat, cholesterol-free, semipurified diet containing the soy protein isolate excreted significantly more neutral and acidic steroids than those fed with casein. As shown in Table 4-I, cholesterol was absorbed to a greater extent on the casein diet compared to the soy protein diet. In rats fed a low-fat, cholesterol-free, semipurified diet, Nagata et al [1982] were also able to demonstrate a moderate but significant reduction of cholesterol absorption on feeding the soy protein isolate rather than casein (Table 4-I). The magnitude of the excretion of intravenously injected $[4-^{14}C]$cholesterol as neutral and acidic steroids was also significantly higher for the plant protein. Figure 4-II shows that, in accordance with the quantitative analyses of fecal steroids, the same amount of radioactivity was excreted in rats fed amino acid mixtures equivalent to the soy protein isolate and casein, and the rate of cholesterol absorption was identical for these two mixtures (Table 4-I).

From these observations, the increased fecal excretion of steroids appears to be responsible for the cholesterol-lowering action of soy protein relative to casein, at least in the animal models. In addition to the decrease in intestinal absorption of cholesterol, the flow of intestinal lymph was found to be significantly lower in rats fed a low-fat, cholesterol-free diet containing the soy pro-

TABLE 4-I. Effect of Dietary Protein on Cholesterol Absorption

Animals	Diets	Cholesterol absorption (%)
Rabbits[a]	Casein	
	Low fat	86 ± 2[b]
	High fat	85 ± 3[b]
	Soy protein	
	Low fat	74 ± 4
	High fat	68 ± 4
Rats[c]	Casein	
	Low fat	72 ± 2[b]
	Soy protein	
	Low fat	63 ± 3
Rats[c]	Casein AA-mix[d]	
	Low fat	33 ± 1
	Soy protein AA-mix	
	Low fat	36 ± 2

[a]Measured after simultaneous oral administration of [³H]β-sitosterol and [¹⁴C]cholesterol [Huff and Carroll, 1980a].
[b]Significantly different from the corresponding soy protein groups.
[c]Measured by the isotope ratio method [Nagata et al, 1982].
[d]Amino acid mixture.

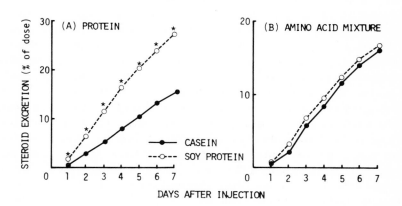

Fig. 4-II. Effect of dietary proteins (A) and amino acid mixtures equivalent to proteins (B) on fecal excretion of the radioactivity as neutral and acidic steroids following intravenous injection of [4-¹⁴C]cholesterol to rats fed low-fat, cholesterol-free, semipurified diets [Nagata et al, 1982]. *Significantly different from the corresponding soy protein group.

tein isolate than in rats fed casein [Sugano et al, 1981]. Though the concentration of lymphatic cholesterol was the same in the two groups of rats, the reduction of the lymphatic flow might accompany the decreased influx of cholesterol from the intestine into circulation via the lymph. It seems likely that the dietary protein type may influence the general functioning of the small intestine.

A significant protein-dependent difference in the proportion of major neutral steroids, but not bile acids, was found in rabbit feces [Huff and Carroll, 1980a]. Thus, rabbits fed the casein diet excreted mainly cholesterol, whereas rabbits fed the soy protein diet excreted coprostanol as the major steroid. This was not the case in the rat experiment [Nagata et al, 1980, 1981a]. Though the results with rabbits suggest a probable protein-dependent alteration in the intestinal flora, the significance of the change in relation to the increased fecal excretion of neutral steroids is obscure. Terpstra and Sanchez-Muniz [1981] recognized in rabbits that soy protein exerted its cholesterol-lowering action within one day (or even in 12 hours) after the ingestion of semipurified diets. The rapidity of the changes induced by feeding of semipurified diets indicates that mechanisms such as alterations in the intestinal flora and transit time are not involved in the hypocholesterolemic action of soy protein, suggesting rather the functioning of an "emergency" mechanism such as hormonal regulation.

Despite these virtually decisive demonstrations of the protein-dependent difference in cholesterol absorption and fecal steroid excretion in the animal models, the results of human studies are inconsistent with these observations. Munoz et al [1979] failed to demonstrate a significant increase in mean daily fecal weight in healthy adult, American men eating an additional 26 g per day of textured soy protein (16% of calories as protein, 70% from animal sources). Noseda and Fragiacomo [1980] could not find a clear-cut change in fecal neutral steroid and/or bile acid output when the proteins in a low-lipid diet were totally replaced with textured soy protein in type II patients (Table 4-II). Potter et al [1980], in their double-blind crossover studies with male normolipidemic volunteers, reported that defatted soy flour essentially free of saponins given at the 15.5% level of total proteins did not increase either fecal neutral or acidic steroids above the corresponding baseline values. The same pattern of response was confirmed with outpatient hypercholesterolemic men [Calvert et al, 1981]. Grundy and Abrams [1981] also studied the cholesterol balance in hyperlipidemic inpatients providing the soy protein isolate at the 15% calorie level in a liquid formula diet. Compared with the casein formula, no marked difference was found in the cholesterol balance or in the excretion of neutral and acidic steroids.

These observations, taken together, suggest that plasma cholesterol-lowering action of soy protein in humans might be exerted by the mechanism appar-

TABLE 4-II. Sterol Balance Studies in Two Type II Patients

	Patient A		Patient B	
	Low lipid	Soybean	Low lipid	Soybean
Total plasma cholesterol (mg/dl)	361	291	400	310
Fecal steroids (mg/day)				
Neutral	730	612	858	726
Acidic	296	243	304	284
Total	1026	855	1162	1010

From Noseda et al [1980].

ently different from that operating in the experimental animals. Sirtori et al [1980] ruled out the hypothesis that in humans the protein effect is simply mediated by a cholesterol or bile acid malabsorption. These differences between species may instead have to do with the difference in the extent of the change in plasma cholesterol levels in response to the type of dietary protein. Thus, there is still a possibility that soy protein exaggerates fecal output of steroids even in humans when diets strictly equivalent to those ingested in the animal experiments are given. In contrast, the reverse may not be true since the soy protein diet, which failed to show the hypocholesterolemic effect in human volunteers, certainly exerted a significant hypocholesterolemic effect in rabbits [van Raaij et al, 1981].

The observation that is instructive as to the effect of protein types on the fecal excretion of steroids is that reported by Potter et al [1976] in infants from the first week of life to 2 or 3 months of age: The substitution of soybean milk (polyunsaturated to saturated fatty acid [P/S] ratio 4.2; cholesterol-free) for cow's milk (P/S ratio 0.04; cholesterol 8 mg/dl) lowered the plasma cholesterol (even in the presence of dietary cholesterol) and led to a twofold increase in bile acid excretion. They also concluded that soybean milk probably increases fecal cholesterol excretion. Of course, the effect of the difference in P/S ratio should be taken into account, since steroid excretion increases in response to an increase in P/S ratio [Grundy, 1975; Nestel et al, 1975]. In rabbits, the type of dietary protein did not alter the hypercholesterolemia during weaning [Roberts et al, 1979]. After weaning, however, animals fed milk protein (casein) in the absence of cholesterol became hypercholesterolemic, whereas those fed the soy protein isolate remained normocholesterolemic.

Alternatively, the responses obtained in human studies could be compatible with those obtained in rats fed an amino acid mixture diet. Though changes in fecal steroid output may not necessarily accompany changes in plasma cholesterol levels [Sautier et al, 1979; Potter et al, 1980; Calvert et al, 1981], the observation that steroid excretion in human subjects remains apparently unchanged in response to the protein type may be connected with the specificity of the cholesterol metabolism in this species relative to rodents. In order to determine whether a mechanism similar to that currently proposed for the animal models could be applicable to humans, further studies under more appropriate protocols are indispensable.

III. EFFECTS ON HEPATIC SYNTHESIS AND SECRETION OF CHOLESTEROL

Owing to a well-known feedback regulation mechanism of hepatic cholesterol synthesis mediated by cholesterol delivered from the intestine, the decreased intestinal absorption of cholesterol should reciprocally cause an increase in cholesterogenesis in that tissue. Reiser et al [1977] in fact demonstrated an increment of the specific activity of 3-hydroxy-3-methyl-glutaryl coenzyme A (HMG-CoA) reductase, the rate-limiting enzyme in the cholesterol synthetic pathway, in the liver of rats fed a semipurified diet containing soy protein rather than casein (Table 4-III). The observation of Mokady and Einav [1978] that a higher rate of incorporation of acetate into liver cholesterol was achieved in rats fed a gluten diet than in those fed a casein diet may also point to the mitigation of the feedback inhibition.

In swine on mash diets to which textured soy protein or casein were added, Kim et al [1980a] were unable to find a change in hepatic HMG-CoA reductase activity related to the two protein types. Based on a similar drop in the reductase activity between swine fed the hyperlipidemic diet containing soy protein or casein as a sole source of protein and swine fed the mash diet [Kim et al, 1980b], they concluded that in swine fed high-fat, high-cholesterol diets, the effect of soy protein on lowering serum cholesterol levels relative to casein appeared to be an increase in fecal steroid excretion not counterbalanced by a concomitant increase in cholesterol synthesis in the liver [Kim et al, 1980a]. When diets high in fat but essentially free of cholesterol (HL diet) were fed to swine [Kim et al, 1980a], whole-body cholesterol synthesis of both HL + casein and HL + soy protein was the same, but whole-body cholesterol concentration was significantly higher in the former diet, though total steroid excretion was comparable in both.

The disproportion concept can in part be understood by the observation that soy protein as compared with casein lowers the cholesterol concentration in the liver of rabbits [Huff and Carroll, 1980a] and rats [Nagata et al, 1981a]. Recently, more direct evidence was introduced by Huff and Carroll [1980a] in rabbits. They reported that oxidation of cholesterol, calculated from the output of $^{14}CO_2$ in expired air following administration of [26-^{14}C]cholesterol, occurred at a faster rate in rabbits on the soy protein diets than in rabbits on the casein diets. Accompanying this modulation was a significant increase in the

TABLE 4-III. Effect of Dietary Protein on Hepatic Activities of HMG-CoA Reductase in Rats Fed Semipurified Diets

Diets	HMG-CoA reductase (nmol/min/mg protein)
I[a] Casein	0.115 ± 0.020[b]
Soy protein	0.247 ± 0.043
II[c] Casein	0.121 ± 0.014[b]
Soy protein	0.253 ± 0.052
Casein AA-mix[d]	0.368 ± 0.052[b]
Soy protein AA-mix	0.195 ± 0.020

[a]Reiser et al [1977].
[b]Significantly different from the corresponding soy protein groups.
[c]Nagata et al [1982].
[d]Amino acid mixture.

metabolic turnover of cholesterol in the rapidly exchangeable pool, pool A [Goodman and Noble, 1968; Nestel et al, 1969]. Experiments with rats fed a similar type of diet also showed that the expansion of the size of pool A due to feeding casein instead of soy protein was a reflection of the decreased removal rate of cholesterol rather than the increased production of cholesterol in that pool [Nagata et al, 1982]. These observations, considered together, suggest that the bulk of cholesterol synthesized in the liver from the rodents fed soy protein is exclusively excreted through bile as bile acids and/or cholesterol rather than secreted into the plasma compartment.

However, uncertainty still exists about whether the liver in rats fed soy protein actually secretes less cholesterol to circulation. This can be estimated appropriately by measuring the rate of lipoprotein secretion by the perfused liver or hepatocytes. We have examined the effect of the dietary protein type on the secretion of lipoproteins by the isolated perfused rat liver [Sugano et al, 1982b]. The results shown in Figure 4-IIIA clearly demonstrate a protein-dependent difference in the rate of lipoprotein secretion: Rats fed a low-fat, cholesterol-free diet containing casein secreted 1.4-fold more cholesterol (also about twice as much triglyceride and apo-A_I) for four hours of perfusions than the animals fed the corresponding soy protein diet. Thus, it is likely that, in addition to the disproportion between a decreased delivery of cholesterol from the intestine and an increased hepatic sterogenesis, the reduction of hepatic contribution potential is directly relevant to the decrease in serum cholesterol levels.

The situation completely differed when amino acid mixture diets simulating casein and soy protein were fed to rats. As stated before, rats fed the soy protein-type amino acid mixture absorbed cholesterol and excreted steroids at the same rate as that seen in the animals fed the casein-type counterpart

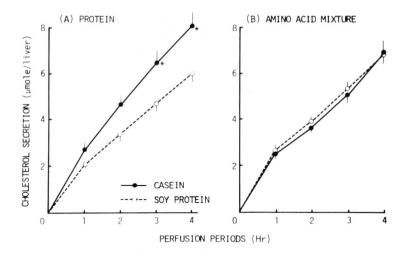

Fig. 4-III. Effect of dietary proteins (A) and amino acid mixtures equivalent to proteins (B) on secretion of cholesterol by isolated perfused rat liver. Livers from rats fed low-fat, cholesterol-free diets were perfused with Krebs-Henseleit buffer containing glucose and washed aged human erythrocytes [Sugano et al, 1982b]. *Significantly different from the soy protein group.

[Nagata et al, 1981a]. In spite of these circumstances, HMG-CoA reductase activity in the liver was significantly higher in the latter group [Nagata et al, 1982] (Table 4-III). Since isolated livers from both dietary groups of rats secreted cholesterol (and also triglyceride and apo-A_I) at essentially the same rate (Fig. 4-IIIB) and since there were no demonstrable differences in concentrations of liver cholesterol and triglyceride levels between two groups [Nagata et al, 1981a], it seems that the increase in the concentration of serum cholesterol due to the casein-type compared with soy protein-type amino acid mixture might in part be merely a reflection of increased hepatic synthesis of cholesterol. In fact, the multicompartmental analysis of the slope of the serum cholesterol specific-activity curve according to the two-pool model revealed that the expansion of the size of pool A on a casein-type mixture resulted from an increased production rate in that pool [Nagata et al, 1982].

Though more detailed studies are necessary before making decisive conclusions concerning the role of the liver in the regulation by dietary protein of serum cholesterol levels, the modification of cholesterol dynamics observed in the liver of rats fed the amino acid mixture diet suggests the possibility that such mechanisms may be operating in humans.

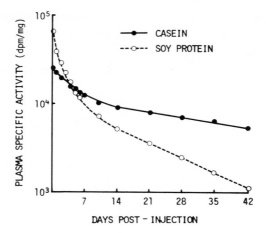

Fig. 4-IV. Effect of dietary proteins on the decay of plasma cholesterol specific activity following injection of [26-¹⁴C]cholesterol to rabbits fed low-fat, cholesterol-free, semipurified diets. The data of Huff and Carroll [1980a] were simplified.

IV. EFFECTS ON CHOLESTEROL TURNOVER

As depicted in Figure 4-IV, the decay of the specific activity of cholesterol in the plasma of rabbits fed the soy protein diet was much more rapid than in rabbits fed the casein diet [Huff and Carroll, 1980a]. The results obtained by analyzing the decay curve according to the two-pool model indicated the following changes in kinetic parameters of cholesterol metabolism in rabbits fed the soy protein diet compared to those fed the casein diet: (1) the reduction in mass of cholesterol accompanying the increase in the metabolic turnover of cholesterol in pool A and (2) the transfer of cholesterol from pool A to pool B (slowly exchangeable pool) and the minimum mass of cholesterol in pool B were both decreased. These changes were all statistically significant. A similar pattern of the difference, though not so marked as in rabbits, was also observed in rats as illustrated in Figure 4-VA [Nagata et al, 1982]. Similarly, feeding soy protein to rats resulted in the decreasing size of pool A and the increasing rate of removal from this pool. There was, however, no marked difference in the removal rate of cholesterol from pool B between rats fed soy protein and those fed casein. As the results indicate, the extent of the deposition of intravenously injected [4-¹⁴C]cholesterol was markedly less in several peripheral tissues including adrenals, testes, adipose tissues, and aorta as well as liver of rats fed soy protein [Nagata et al, 1982].

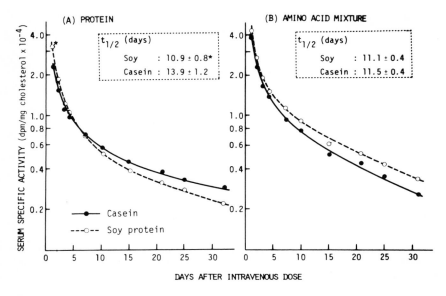

Fig. 4-V. Effect of dietary proteins (A) and amino acid mixtures equivalent to proteins (B) on the decay of serum cholesterol specific activity following injection of [4-^{14}C]cholesterol to rats fed low-fat, semipurified diets. *Significantly different from the casein group. [Nagata et al, 1982].

The increase in fecal excretion of steroids as a reflection of the reduction of their absorption, the increased oxidation of cholesterol in the liver, and the decreased secretion of cholesterol by the liver in combination point to the mechanism responsible for the hypocholesterolemic effect of soy protein (compared with casein) in rabbits and rats.

Interestingly, when casein or soy protein in a semipurified diet was replaced with the corresponding amino acid mixture, the contours of the serum cholesterol decay curve of rats became essentially the same (Fig. 4-VB). Thus, the mechanism for the cholesterol-lowering action indicated in the rodents fed intact soy protein diets could not be adapted when the amino acid mixture simulating soy protein was fed to rats. It is, however, unclear whether the mechanism by which the soy protein-type amino acid mixture exerts the anti-hypercholesterolemic effect is operating simultaneously in the intact soy protein.

The studies on the effect of the protein type on plasma lipoprotein metabolism in rabbits will be discussed later (section VII).

TABLE 4-IV. Effect of Enzymatic Hydrolysates and Amino Acid Mixtures on Plasma Cholesterol Levels in Rabbits and Rats

Animals	Diets	Plasma cholesterol (mg/dl)
Rabbits[a]	Casein	
	Intact	213 ± 53[b]
	Enzymatic hydrolysate	178 ± 30[b]
	Amino acid mixture	213 ± 42[b]
	Soy protein	
	Intact	69 ± 12
	Enzymatic hydrolysate	41 ± 8
	Amino acid mixture	124 ± 30
Rats[c]	Casein	
	Intact	87 ± 5[d]
	Amino acid mixture	87 ± 3[d]
	Soy protein	
	Intact	69 ± 5
	Amino acid mixture	73 ± 6

[a]Rabbits were fed these diets for 28 days [Huff et al, 1977].
[b]Significantly different from intact soy protein.
[c]Rats were fed these diets for 21 days [Nagata et al, 1981a].
[d]Significantly different from the soy protein groups.

V. EFFECTS OF AMINO ACID COMPOSITION

On continuing the investigation of hypocholesterolemic action of the plant proteins, it seems logical to examine the effects of differences in amino acid composition on plasma cholesterol levels. This is also relevant to whether the protein component itself is actually responsible for the regulation of plasma cholesterol levels.

Based on a considerable difference in the amino acid profile between the soy protein isolate and casein, it has long been thought that the difference in the amino acid composition should account for different effects of dietary proteins on plasma cholesterol. A number of studies along these lines have been carried out.

A. Intact Protein Versus Amino Acid Mixture

Carroll and his co-workers [Huff et al, 1977] examined the effect of substituting enzymatic hydrolysates of proteins or amino acid mixtures simulating proteins for the intact proteins on the plasma cholesterol level of rabbits fed low-fat, cholesterol-free diets. The results are summarized in Table 4-IV together with those obtained in a similar experiment with rats [Nagata et al, 1981a]. An enzymatic hydrolysate of casein or a mixture of L-amino acids equivalent to casein gave elevated plasma cholesterol levels similar to those obtained with the intact protein. Though an enzymatic hydrolysate of soy protein also gave cholesterol levels similar to those obtained with the intact protein, a moderate, but not significant, increase in plasma cholesterol was observed when a mixture of L-amino acids equivalent to the soy protein isolate was substituted for the intact protein; and as a result of the increase, the difference between casein-type and soy protein-type mixtures became less clear. Huff et al [1977] concluded from these observations that the level of plasma cholesterol could be influenced by the amino acids supplied in the diet. Thus, the results in fact supported the concept that the hypercholesterolemia obtained with the casein-containing diet was not due to any nonprotein components of the protein preparation, while it was less clear whether the ability of soy protein to maintain lower plasma cholesterol levels could be attributed to the difference in amino acid composition alone. Since the pancreatin was used to prepare the hydrolysates, the bulk of amino acids should be present in the peptide form in the protein hydrolysates. Their observations strongly suggest that some kinds of hydrolysates with a peptide nature may play a crucial role in regulating the concentration of plasma cholesterol (refer to section VI).

Yadav and Liener [1977] studied in rats the effect of amino acid mixtures equivalent to soy protein and casein in diets rich in both cholesterol and fat. Though they presented a clear difference in plasma cholesterol levels between casein- and soy protein-type mixtures, the evaluation of their data was subject to restriction since the diets were virtually depleted of essential fatty acids. Subsequently, Nagata et al [1981a] verified in rats (by feeding a low-fat, cholesterol-free diet) that the serum cholesterol-lowering action of soy protein could be duplicated even when that protein was replaced with the corresponding amino acid mixture (Table 4-IV). No sign of essential fatty acid deficiency as judged by the spectra of fatty acid compositions in various tissues was observed under their conditions. Nagata et al [1981b] also confirmed that serum cholesterol levels were significantly lower in the soy protein-type amino acid mixture than in the casein-type counterpart even when cholesterol was added to a low-fat diet, though it required ten weeks to demonstrate a clear-cut difference (see Fig. 4-VII). These observations favor the view that the difference in the amino acid composition is responsible for most, if not all, of the difference in plasma cholesterol levels of the rodents, particularly in rats. However, this tentative conclusion should be evaluated with considerable caution because of the notable difference in the sequence of intestinal absorption between the intact protein and amino acid mixture [Sleisenger and Kim, 1979; Silk, 1980].

The soy protein isolate is composed mainly of different kinds of globulins. Okita and Sugano [1981] studied what types of globulins are being preferentially involved in the hypocholesterolemic effect of soy protein in rats. They prepared three globulin fractions: 11S, 7S, and total globulins. There were slight but detectable differences in amino acid composition among these preparations. In connection with the following argument, the ratio of lysine/arginine (Lys/Arg) was lower in 11S than in 7S globulins. In a low-fat, cholesterol-free diet 11S globulin tended to produce a lower level of plasma cholesterol than the 7S counterpart. The results also indicated that the antihypercholesterolemic effect of soy protein was attributable to the protein component itself and not to the carbohydrate components which associated with the protein preparations.

B. Methionine (Met) Contents

The soy protein isolate contains about one-half Met relative to casein. The effect of addition of Met to soy protein, therefore, attracted attention. Sufficient amounts of Met should, however, certainly be provided if dietary protein levels are maintained at around 20% as usually applied for the studies conducted so far. Since rats and rabbits given soy protein appeared to grow normally relative to the animals given casein, the hypocholesterolemic effect of soy protein will presumably be discriminated from the function of Met as an essential nutrient [Leiner, 1977].

In fact, Huff et al [1977] reported that the supplementation of Met to soy protein modified only slightly the level of plasma cholesterol of rabbits, indicating that this essential amino acid may not positively be involved in the control of cholesterol levels. The study of Okita and Sugano [1981] also favors this concept: The plasma cholesterol levels of rats fed 11S globulin were lower than those fed 7S globulin; the former contained three times as much Met as the latter. It seems again, therefore, that Met may not be a major contributing factor in the hypocholesterolemic action of soy protein.

C. Lysine/Arginine (Lys/Arg) Ratio

Differences in compositions involving amino acids other than Met have also drawn attention. Among these, the Lys/Arg ratio hypothesis seems most attractive. Kritchevsky [1979] considered the Lys/Arg ratio as a determining factor on the basis of the antagonism between these two basic amino acids: Lysine interferes with arginase and hence the higher Lys/Arg ratio may cause more Arg to be available to be incorporated into "atherogenic" arginine-rich apoproteins (apo E) [Shore et al, 1974a,b]. The ratio is twofold higher in casein. Thus, Kritchevsky and co-worker [Kritchevsky, 1979; Czarnecki and Kritchevsky, 1980] observed in rabbits that supplementation of Lys to soy pro-

TABLE 4-V. Effect of Lysine and Arginine Addition to Soy Protein and Casein on Serum Cholesterol and Experimental Atherosclerosis in Rabbits

Diets[a]	Serum cholesterol (mg/dl)	Average atheromata[b]	
		Arch	Thoracic
Experiment I			
Casein	174 ± 30	2.2 ± 0.5	1.5 ± 0.4
Soy protein	59 ± 14	0.8 ± 0.4	0.5 ± 0.2
Casein + Arg	129 ± 12	1.4 ± 0.4	0.8 ± 0.3
Soy protein + Lys	106 ± 29	1.6 ± 0.4	1.1 ± 0.2
Experiment II			
Casein	214 ± 27	1.1 ± 0.3	0.8 ± 0.3
Soy protein	171 ± 18	0.5 ± 0.2	0.4 ± 0.1
Casein + Arg	281 ± 63	1.3 ± 0.4	0.9 ± 0.3
Soy protein + Lys	197 ± 16	0.7 ± 0.2	0.5 ± 0.2

[a]Arginine added to give diet arginine/lysine ratio equal to soy protein. Lysine added to give arginine/lysine ratio equal to casein.
[b]Aortas graded on a 0–4 scale.
From Kritchevsky [1979].

tein to elevate the Lys/Arg ratio to be equivalent to that of casein resulted in the rise of the plasma cholesterol level and the degree of atherosclerosis (Table 4-V). The effect of addition of Arg to casein was not unequivocal. In contrast, Huff and Carroll [1980b] could not demonstrate such an effect of Lys addition to soy protein in the same animal species. Our observation with rats also consistently showed no supplementary effects of these basic amino acids on serum cholesterol [Nagata et al, 1981a] (see also Table 4-VII).

More recently, Liepa and Park [1981] examined in rats the effect of addition of adequate Arg to casein to make the Lys/Arg ratio similar to that found in cottonseed protein and enough Lys to cottonseed protein to make it have an Lys/Arg ratio identical to that of casein. They suggested that the effect dietary protein exerted on serum total and high-density lipoprotein (HDL) cholesterol might be due to the ratios of Lys and Arg within the diets. It is therefore likely that in their study the ratio of Lys/Arg is one of the determinants to the plasma cholesterol level.

In spite of these discrepancies, there is additional evidence supporting the view that the ratio of Lys/Arg in dietary proteins is responsible for regulating the plasma cholesterol level. Kritchevsky and Klurfeld [1979] compared effects of different proteins with different Lys/Arg ratios on plasma cholesterol levels in rabbits. The proteins used were corn protein, wheat gluten, and lactalbumin, and the ratios of Lys/Arg for these proteins were 0.56, 0.43, and 2.63, respectively. The concentration of serum cholesterol was markedly higher in rabbits fed lactalbumin than those fed plant proteins. The differ-

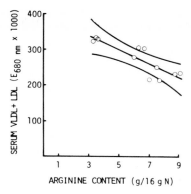

Fig. 4-VI.   Regression of serum VLDL plus LDL on dietary arginine content in experiments with female rats fed different types of dietary protein, with 99%-confidence belt for the prediction of the population regression line [Eklund and Sjöblom, 1980].

ences in atherogenesis were also evident. Hevia et al [1980b] also demonstrated in rats the possible effect of Lys-Arg antagonism on lipid metabolism.

Though Kritchevsky's hypothesis is indeed attractive, it is quite difficult to explicate why soy protein exhibits the hypocholesterolemic effect in spite of its high content of Arg relative to casein. Totally unclear is the question of whether Lys-Arg antagonism is more important for the regulation of plasma cholesterol than the actual amount of dietary Arg. Thus, at present, the relative importance of Arg of endogenous and exogenous origin remains obscure. The effect of Lys/Arg ratio in diets will be examined further in section VI.

D. Lysine or Arginine Contents

Another possibility is that the actual level of Lys or Arg is the determining factor rather than the ratio of the two. Though the studies of Weigensberg et al [1964] indicated that the diets high in Lys were hypercholesterolemic for rabbits, Huff and Carroll [1980b] were unable to show a strong correlation of dietary Lys with plasma cholesterol levels. Eklund and Sjöblom [1980] reported a significant correlation between dietary Arg content and plasma very low density lipoprotein (VLDL) and low density lipoprotein (LDL) levels in female rats as shown in Figure 4-VI. Since the Lys content of soy protein is comparable with that of casein, it is reasonable to consider that it is the Arg that has a crucial role in the regulation of plasma cholesterol [Hevia et al, 1980a].

E. Glycine (Gly) Contents

One of other amino acids which presents in different amounts is Gly. Hermus and Dallinga-Thie [1979] failed to find effects of increasing the Lys/

Arg ratio on serum cholesterol of rabbits, but in contrast they provided evidence that supplementation of casein diets with Gly might depress the serum cholesterol to a lesser extent. Subsequently, Katan et al [1980] observed that casein plus 0.55% Alanine (Ala) and 0.35% Arg (CasAA) or CasAA plus 0.65%, 1.40%, or 3.15% Gly in the 20% casein diet reduced not only mean serum cholesterol levels but also the degree of atherosclerosis in rabbits. The observation may be relevant to the role of Gly as a probable semiessential amino acid for rabbits. The Gly effect, however, appeared to be nonspecific, since in rats addition of 2.0% Gly to the 20% casein diet rich in cholesterol similarly reduced serum cholesterol level to that of the animals fed soy protein. Aust et al [1980] reported the hypolipemic action of a glycine-rich diet in rats. It is unclear if the observations can be extrapolated to humans.

F. Proportion of Nonessential to Essential Amino Acids

Detailed and systemic studies made by Huff and Carroll [1980b] indicated that the different effects of soy protein and casein were not attributable to a single amino acid but rather to the proportion of essential to nonessential amino acids. The dietary ratio of Σisoleucine, tyrosine, threonine, serine/ ΣArg, Gly, and glutamic acid was important in regulating plasma cholesterol levels in rabbits fed a low-fat, cholesterol-free, semipurified diet. When protein components were formulated by adding amino acids to casein to give a mixture corresponding to soy protein or by adding amino acids to soy protein to give a mixture equivalent to casein, these diets failed to reserve plasma cholesterol levels. These observations led the authors to consider that influences of differences in digestion and absorption of proteins relative to amino acids are by no means to be ignored [Sleisenger and Kim, 1979; Silk, 1980].

G. Relation to Plasma Amino Acid Pattern

The more important point to be clarified with respect to the effect of amino acid composition on the plasma level of cholesterol observed in experimental animals is whether these dietary effects can be extrapolated to humans. Most of the human data have been derived from the relatively short-term feeding trials. Thus, the effect of soy protein on plasma cholesterol levels in human subjects, in particular normocholesterolemic volunteers [Carroll et al, 1978; van Raaij et al, 1981] or mild hypercholesterolemics [Holmes et al, 1980; Grundy and Abrams, 1981; Shorey et al, 1981], is rather moderate or sometimes negligible. For confirmation of the protein effect, long-term experiments seem prerequisite. In addition, only limited information is available as to the effect of addition of specific amino acid(s) to diets on plasma cholesterol levels.

Jarowski and co-workers showed that Lys (and tryptophan [Trp]) supplementation lowered plasma cholesterol not only in rats [Jarowski and Pytelewski, 1975] but also in hyperlipidemic subjects [Raja and Jarowski, 1975], an observation which contradicts the hypothesis of Kritchevsky [1979]. Since Jarowski's diets were either commercial rat chow or customary American diets, the comparison of the results with those obtained with more formulated diets should be subject to a respectable restriction. They subsequently investigated to determine if the reverse situation is the case [Torre et al, 1980]. Rats fed laboratory rations with added gelatin (deficient in Trp) showed significant increases in serum cholesterol and triglyceride levels, while supplementation of Trp to these diets obviated the hyperlipidemia.

The idea of Jarowski's studies stemmed from the comparison of differences in amino acid composition in diets and fasting blood plasma [Jarowski and Pytelewski, 1975]. Nagata et al [1980] found no differences in the fasting serum amino acid profile in rats fed low-fat, cholesterol-free, semipurified diets containing either the soy protein isolate or casein, but there was a protein-independent change in the profile when cholesterol was included in these diets. The commonly observed change was a significant reduction of the concentration of serum threonine (Thr), suggesting that the specific amino acid(s) participates preferentially in the metabolism of cholesterol. In swine fed a high-cholesterol, semipurified diet, plasma levels of valine (Val), isoleucine (Ile), and leucine (Leu) in addition to Thr were significantly lower for plant proteins than for animal proteins [Forsythe et al, 1980].

Garlich et al [1970] were unable to explain the relationship between the plasma free amino acid concentration and the hypocholesterolemic action in adult humans. Olson et al [1970] observed marked hypocholesterolemia in human subjects fed an amino acid formula diet containing the eight essential amino acids required by adult humans plus glutamic acid as the sole source of nonessential amino nitrogen. Bazzano et al [1970] also found a reduction of serum cholesterol and associated beta-lipoproteins by oral administration of large amounts of Glu to adult humans in a similar formula diet. The mechanism of the hypocholesterolemic action of these diets is unknown.

In contrast, Anderson et al [1971] failed to show any difference in the concentration of serum cholesterol or triglyceride when healthy male students were fed two diets differing only in the composition of 60 g of daily protein (half the total protein intake), viz, wheat gluten or egg white. As compared with egg white, gluten is lower in aspartic acid (Asp), Lys, Met, and alanine (Ala) but much higher in Glu and proline (Pro); thus the gluten diet contained about 20 g more Glu daily.

VI. EFFECTS ON HORMONAL STATUS

The available evidence shows that the plasma cholesterol level in rabbits [Huff et al, 1977; Huff and Carroll, 1980b] and rats [Nagata et al, 1981a] can

TABLE 4-VI. Plasma Glucagon, Insulin, and Glucagon: Insulin Ratio Before, During, and After Treatment With Soybean Proteins in Type II Patients

Hormone (pg/ml)	Before	During (weeks)		After (6 weeks)
		4	8	
Glucagon	218 ± 114	252 ± 98	247 ± 95[a]	215 ± 101
Insulin	19.9 ± 5.3	19.2 ± 4.7	16.4 ± 4.9	18.7 ± 6.5
Glucagon insulin ratio	11.5 ± 6.3	13.9 ± 7.1	15.7 ± 6.8[a]	11.8 ± 5.2

[a]Significantly different from the corresponding before-values.
From Noseda and Fragiacomo [1980].

be modified by the amino acid composition of dietary proteins. The amino acid mixture appears to affect the plasma cholesterol level in a manner different from that of the intact protein. As suggested by Huff and Carroll [1980b] and as mentioned previously, the influence of the structure of the dietary proteins, namely, amino acid sequence, seems to be of importance.

As discussed in the preceding section, Arg appears one of the most important amino acids in connection with the regulation of the concentration of plasma cholesterol. It has been known that Arg, among amino acids, is the most effective stimulus both for insulin [Fajans et al, 1968] and glucagon [Assam et al, 1977] secretion. Thus, the relative and absolute abundance of Arg in soy protein as compared with casein may be relevant to the antihypercholesterolemic effect of that protein. The interaction between dietary proteins and hormonal status was first demonstrated by Noseda and Fragiacomo [1980]. They showed in hypercholesterolemic patients that when animal proteins in their customary diets were replaced with textured soy protein, plasma levels of glucagon increased significantly in eight weeks, while insulin remained apparently unchanged, resulting in an elevation of the glucagon/insulin ratio (Table 4-VI).

Noseda et al [1980] simultaneously observed that the magnitude of the rise of the plasma glucagon level after infusion of Arg was also significantly greater after feeding a soy protein diet for one week. There was a selective increase in the small molecular weight glucagon form with a corresponding decrease in the large molecular weight form of the circulating hormone. The low molecular weight glucagon is supposed to be certainly effective on intermediary metabolism in contrast to the "big hormone" secreted by the gut [Srikant et al, 1977]. Further studies are needed to evaluate the significance of this observation. Kritchevsky [1979] hypothesized that in animals given casein more Arg might be available to be incorporated into apo-E as a reflection of the probable depression of arginase activity by Lys [Cittadini et al, 1964]. Alternatively, this circumstance is likely to stimulate the secretion of glucagon or insulin.

TABLE 4-VII. Effect of Lysine and Arginine Addition to Soy Protein and Casein on Serum Lipids and Plasma Insulin and Glucagon Levels in Rats

Diets (Lys/Arg ratio)[a]	Serum lipids (mg/dl)		Plasma hormones	
	Cholesterol	Triglyceride	Insulin[b] (μU/ml)	Glucagon (pg/ml)
Experiment I				
Soy protein (0.93)	76 ± 7[c]	171 ± 27[c,d]	113 ± 15[c]	108 ± 9[c]
Casein (2.09)	125 ± 11[d]	195 ± 27[c,e]	188 ± 18[d]	112 ± 8[c]
Casein + 1 Arg (0.93)	128 ± 20[d]	239 ± 9[e]	218 ± 39[d]	152 ± 22[c,d]
Casein + 2 Arg (0.60)	126 ± 10[d]	166 ± 21[c,d]	249 ± 32[d]	146 ± 12[d]
Casein + 5 Arg (0.29)	122 ± 12[d]	117 ± 19[d]	161 ± 27[c,d]	167 ± 14[d]
Experiment II				
Casein (2.09)	100 ± 6[c]	141 ± 15[c]	63 ± 3[c]	101 ± 9[c]
Soy protein (0.93)	69 ± 3[d]	92 ± 8[d]	47 ± 2[d]	113 ± 14[c,d]
Soy protein + 1/2 Lys (1.51)	72 ± 3[d]	153 ± 21[d]	70 ± 6[c]	154 ± 20[d]
Soy protein + 1 Lys (2.09)	63 ± 4[d]	150 ± 13[c]	52 ± 6[c,d]	107 ± 8[c,d]
Soy protein + 2 Lys (3.24)	71 ± 5[d]	188 ± 20[c]	78 ± 14[c,d]	122 ± 10[c,d]

[a]Rats were fed these diets for 40 days.

[b]Due to the methodological difference in the determination of immunoreactive insulin between experiments I and II, the absolute values differed.

[c,d,e]In each experiment, values in the same vertical column not sharing common superscript letters were significantly different.

From Sugano et al [1982a].

This possibility was studied by widely modifying the ratio of Lys/Arg by adding Arg to casein and Lys to the soy protein isolate [Sugano et al, 1982a]. As summarized in Table 4-VII, in rats fed a low-fat, cholesterol-free, semi-purified diet containing protein at the 20% level, either the addition of Arg to casein or Lys to soy protein did not modify the intrinsic ability of these proteins to regulate serum cholesterol levels. Thus, in unmodified (Lys/Arg ratio 2.09) and modified (Lys/Arg ratios 0.93, 0.60, and 0.29, respectively) casein diets, concentrations of serum cholesterol all ranged in the higher levels relative to those of the soy isolate diet. Simultaneously, in unmodified (Lys/Arg ratio 0.93) and modified (Lys/Arg ratios 1.51, 2.09, and 3.24, respectively) soy protein diets, serum cholesterol remained at the same lower levels relative to the unmodified casein diet. In contrast to the steadiness of serum cholesterol levels, there was an apparent dose-dependent change in the concentration of serum triglyceride; it tended to decrease accompanying a decrease in the Lys/Arg ratio caused by adding Arg to casein and increase on increasing the ratio by adding Lys to soy protein.

Casein caused an increase in the concentration of immunoreactive insulin in rat serum, whereas that of plasma glucagon (30K-antibody) remained unchanged [Sugano et al, 1982a] (Table 4-VII). The glucagon level increased proportionately with increasing amounts of Arg supplemented to casein, while Lys addition to soy protein showed no such an effect. Supplementary effect of these amino acids on serum insulin levels were not unequivocal. In addition, there was a parallel increase in the concentration of serum apo-E and plasma glucagon in response to Arg supplementation of casein. Addition of Lys to soy protein also increased apo-E, though independently on the dose level.

Increased insulin secretion has been demonstrated to produce increased hepatic production of triglyceride [Reaven et al, 1967] and cholesterol [Dugan et al, 1974] followed by stimulation of triglyceride secretion as VLDL. Therefore, the elevation of serum cholesterol and triglyceride in rats given casein as compared with those given soy protein appears to be at least governed by the increased insulin secretion. The activity of lipoprotein lipase is also under control of insulin and/or glucagon [Nilsson-Ehle, 1981]. A dose-dependent increase in plasma glucagon with a concomitant decrease in serum triglyceride when Arg was added to casein falls into line with a potent hypolipidemic effect of that hormone as a reflection of the reduced hepatic synthesis and subsequent secretion of triglyceride-rich lipoproteins [Eaton, 1973; Schade et al, 1979]. No such a correlation was, however, found between blood levels of glucagon and triglyceride when Lys was added to soy protein, though the triglyceride level increased with increasing amounts of supplementary Lys. In any case, the ratio of Lys/Arg was associated more intimately with regulation of serum triglyceride than cholesterol levels. Insulin appeared to be closely related to different effects of these proteins on serum lipids of rats, though this was not the case in hyperlipidemic patients where glucagon presumably plays a more important role in regulating plasma cholesterol levels [Noseda and Fragiacomo, 1980; Noseda et al, 1980].

There remains the question of whether supplementary free amino acids behave in the same way as those occurring as peptides in the course of intestinal digestion and absorption. Recent knowledge on the digestion and absorption of dietary proteins clearly established marked differences in these parameters depending on whether the dietary nitrogen sources were given as free amino acids or as peptides [Sleisenger and Kim, 1979; Silk, 1980]. In that respect, Eisenstein et al [1979] reported that amino acids or peptides liberated during protein digestion function as glucagon secretagogues through some mechanism other than increased blood amino acid levels after protein ingestion; protein intake itself was supposed to influence alpha-cell function by stimulating release of gastrointestinal hormones. It is therefore necessary to examine whether differences in Arg contents in dietary proteins are actually responsible for the plasma level of glucagon and hence the concentration of plasma lipids including cholesterol and triglyceride.

Striking hormonal responses observed after feeding different types of proteins for relatively short terms may rule out the hypothesis that the effect of plant protein is simply mediated by a cholesterol and/or bile acid malabsorption [Sirtori et al, 1980]. There is a possibility that some components of a peptide nature which influence the secretion of hormones, in particular insulin or glucagon, may be produced specifically during the digestion of soy protein. Alternatively, as suggested by Sirtori et al [1980], it may be that, by analogy with the bioavailability problem encountered in clinical pharmacology, soy protein is properly taken up by the intestinal wall, later releasing some components of a protein nature which lower the serum cholesterol levels.

Our perfusion studies with isolated rat livers suggested a difference in intra-mitochondrial availability of reduced pyridine nucleotide (NADH) between rats fed soy protein and those fed casein [Sugano et al, 1982b]. Thus the production rate of total ketone bodies was the same between rats fed different protein types, but the ratio of β-hydroxybutyrate to acetoacetate was consistently higher on feeding the plant protein. The extent of the glucose secretion into the perfusate was higher in rats fed soy protein than those fed casein, indicating several changes in the balance of glucose metabolism such as glucogenolysis, gluconeogenesis, or glycolysis. From these observations, it may also be presumed that the metabolic sequence for energy production is modified by the type of dietary protein, suggesting a connection of insulin or glucagon with the regulation of serum lipid levels.

The other possible participant that may cause a change in the hormonal secretory potential of the pancreas is the trypsin inhibitor present in the soy protein preparations. Though the inhibitor is usually destroyed by heating during processing the products, it is practically impossible to exclude all the inhibitor activity completely. Trypsin inhibitor acts via the secretory intermediary, cholecystokinin, to stimulate gallbladder contraction and bile output as well as pancreatic enzyme secretion. Long-term feeding of raw soy flour has been found to create pancreatic hypertrophy and hyperplastic and denomatous nodules in rats [McGuiness et al, 1980]. Thus, we may not be able to entirely exclude the possibility that the remaining activity still exerts some influence on the pancreatic functioning to secrete the hormones. The study of Roy and Schneeman [1981] in mice, however, showed that soybean trypsin inhibitor did not seem to affect cholesterol metabolism, though it greatly affects pancreatic enzyme secretion. No evidence of an increase in circulating antibodies of any classes (IgG, IgA, and IgM) to dietary soybean protein could be found in young healthy volunteers in spite of substantial and sustained increase in the intake of various forms of soy protein products [Goulding et al, 1980].

Additional hormonal factors which may participate in the cholesterol-lowering action of soy protein are thyroxine and estrogenic activity. The presence of goitrogenic substance in soy protein are thyroxine and estrogenic activity. The presence of goitrogenic substance in soybean has been reported by several investigators, and the goitrogens were identified to be saponins and

isoflavonoids [Kimura et al, 1979]. Since the thyroid hormone plays a significant role in the metabolism of cholesterol, its function should also be taken
into account, though the heat treatment repressed the hypertrophic potential
to some extent. Soy protein was found to be a better protein supplement than
casein for its hypolipidemic effect because it also decreased liver triglyceride in
the hyperlipidemic hypothyroid chick model [Raheja and Linscheer, 1980].
Both soybean meals used in the manufacture of animal feed and soybean
products intended for human consumption contain genestein and daidzein,
which are found to be estrogenic in mice [Drane et al, 1980]. Even though
these are present at low levels, the effect of estrogenic activity on the regulation
of plasma lipid levels may not be overlooked.

VII. EFFECTS ON SERUM APOLIPOPROTEINS

The information regarding the response of serum apolipoproteins to dietary
proteins is indeed scarce. Almost all of the preceding studies deal only with the
cholesterol level in plasma or in plasma lipoproteins. Since there are striking
species-to-species differences in the distribution of cholesterol among different lipoprotein families, the extrapolation of animal data to humans is under
strict restriction. In view of the peculiar role of apolipoproteins on the metabolism of plasma lipids [Smith et al, 1978; Schaefer et al, 1978], studies aimed to
show the effect of dietary proteins on plasma apolipoproteins should be of
enormous value.

The relatively short-term (21 days) feeding experiments with rats showed
[Nagata et al, 1980, 1981a] that, in a low-fat, cholesterol-free, semipurified
diet, soy protein isolate as compared with casein caused a significant reduction
of serum apo-A_I and the elevation of serum apo-B, while the response of
apo-E was not unequivocal (Table 4-VIII). Similar patterns of responses were
also obtained even when the intact proteins were replaced with the corresponding amino acid mixtures, with the exception of a slight decrease in serum apo-E
on feeding the soy protein-type mixture. Addition of Lys to soy protein or Arg
to casein did not modify these response patterns, and the result was the same
when proteins were replaced with the corresponding amino acid mixtures. The
levels of apo-E in plasma VLDL and LDL were higher in casein-fed rabbits
than those fed soy protein [Carroll, 1981].

As shown in Table 4-VIII, when the feeding period was extended to 40 days
under the same dietary regimen, a similar reduction of serum apo-A_I was obtained in rats fed soy protein compared with rats fed casein, while the level of
apo-B became indistinguishable between plant and animal proteins [Sugano et
al, 1982a].

The reason for changes in the profile of serum apolipoproteins due to the
dietary protein type is not completely clear. In the rats fed the soy protein iso-

TABLE 4-VIII. Effect of Dietary Protein and Amino Acid Mixture on Serum Apolipoprotein Levels in Rats[†]

Diets	Apolipoproteins (μg/ml)		
	Apo-A$_I$	Apo-B	Apo-E
Experiment I[a]			
Casein	1037 ± 122[c]	101 ± 10[c]	239 ± 19[c]
Soy protein	602 ± 29[d]	239 ± 19[d]	245 ± 27[c]
Experiment II[a]			
Casein AA-mix (amino acid mixture)	1091 ± 75[c]	112 ± 12[c]	412 ± 24[c]
Soy protein AA-mix	695 ± 35[d]	139 ± 13[c]	321 ± 30[d]
Experiment III[b]			
Soy protein	468 ± 44[c]	79 ± 9[c]	377 ± 22[c]
Casein	832 ± 69[d,e]	77 ± 9[c]	504 ± 28[d]
Casein + 1 Arg	780 ± 36[d,e]	82 ± 14[c]	495 ± 31[d]
Casein + 2 Arg	821 ± 53[d]	55 ± 7[c]	607 ± 46[d,e]
Casein + 5 Arg	658 ± 42[e]	75 ± 7[c]	644 ± 38[e]
Experiment IV[b]			
Casein	1054 ± 27[c]	103 ± 8[c]	330 ± 16[c,d]
Soy protein	735 ± 45[d]	87 ± 7[c,d]	300 ± 7[d]
Soy protein + 1/2 Lys	729 ± 45[d]	79 ± 4[d]	362 ± 7[c]
Soy protein + 1 Lys	671 ± 87[d]	104 ± 15[c,d]	359 ± 10[c]
Soy protein + 2 Lys	796 ± 109[d]	90 ± 11[c,d]	368 ± 6[c]

[†]See footnotes to Tables 4-VI and 4-VII.
[a]Nagata et al [1981a].
[b]Sugano et al [1982a].
[c,d,e]In each experiment values in the same vertical column not sharing the common superscript letters were significantly different.

late rather than casein, less apo-A$_I$ was secreted by the isolated perfused liver [Sugano et al, 1982b]. Also, though there was no demonstrable difference in the concentration of apo-A$_I$ in the intestinal lymph between rats fed these two proteins, the lymphatic flow rate was significantly lower on feeding soy protein [Sugano et al, 1981]. The significance of the intestine as a contributor of the circulating apo-A$_I$ is frequently indicated [Glickman, 1980; Havel, 1980]. A plausible explanation is therefore that the supply of apo-A$_I$ from the liver and possibly from the intestine is reduced in rats fed the plant protein instead of animal protein. Of course, one cannot exclude the possible effect of the increased turnover of cholesterol on the reduction of serum apo-A$_I$ [Huff and Carroll, 1980a; Nagata et al, 1982]. Forsythe et al [1980] observed in swine fed a high-cholesterol, semipurified diet that feeding animal protein decreased plasma molar lecithin:cholesterol acyltransferase (LCAT) activity (expressed as μmole/liter/hr) and increased plasma unesterified cholesterol levels as compared with feeding plant protein, suggesting alteration of the metabolism of apo-A$_I$, an activator of LCAT. In contrast, LCAT activity was unaltered in rats receiving the soy protein and casein in a hypercholesterolemic diet [Bosisio et al, 1980].

The reason for the transit increase in serum apo-B due to feeding soy protein is also obscure at present. Although the response is not necessarily consistent with the decreasing tendency of the serum triglyceride level in rats fed soy protein, dissociation of apo-B and triglyceride production in VLDL as observed in humans given high-carbohydrate diets [Melish et al, 1980] may also be operating in rats. In any case, a significant difference in the fate of serum VLDL apo-B between rats and humans should be taken into consideration [Smith et al, 1978; Schaefer et al, 1978].

The rise of serum apo-E due to feeding casein to rabbits [Carroll, 1981] and rats [Sugano et al, 1982a] is also not completely explicable, but because of the hypercholesterolemic and atherogenic propensity of this apolipoprotein in rabbits [Shore et al, 1974a,b], the connection to increased atherogenesis can be understood relatively easily. In this context, the regulation of plasma cholesterol by lipoprotein receptors should be taken into account [Brown et al, 1981].

Nagata et al [1981b] observed that when diets rich in cholesterol were fed to rats for ten weeks, both serum apo-A$_I$ and apo-B were reduced on feeding soy protein isolate relative to casein (Fig. 4-VII). Despite these changes, the relative concentration of HDL cholesterol was kept at a higher level in rats fed the plant protein, therefore the increase in serum cholesterol due to casein was attributed to that in VLDL plus LDL cholesterol. These response patterns in the distribution of cholesterol in lipoproteins are comparable with those obtained in rabbits fed low-fat [Carroll et al, 1977] or high-fat [Lacomb and Nibbelink, 1980; Kritchevsky, 1981], cholesterol-free diets. The concentration of serum apo-A$_I$ was similarly decreased even when the proteins were replaced with the corresponding amino acid mixtures in a hypercholesterolemic diet [Nagata et al, 1981b]. Bosisio et al [1980] has reported that apoprotein changes on feeding hypercholesterolemic diets containing either soy protein or casein were minimal.

Few studies are available concerning the effect of dietary soy protein on serum apolipoprotein levels in humans. Richter et al [1980] in a recently published abstract indicated that a reduction of atherogenic LDL/HDL cholesterol ratio by soy protein did not necessarily mean a change of the distribution of the main apolipoproteins-B and -A$_I$ of hypercholesterolemic patients. The results are, however, difficult to evaluate since they administered pectin (3 g/day) simultaneously.

In young, healthy volunteers given diets containing 13% of energy as protein for six weeks, 65% of which consisted of casein or soy protein, van Raaij et al [1981] observed a significantly greater decline in serum apo-B, compared

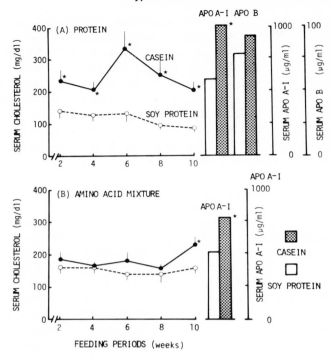

Fig. 4-VII. Effect of dietary proteins (A) and amino acid mixtures equivalent to proteins (B) on serum cholesterol and apolipoprotein levels in rats fed low-fat, cholesterol-enriched diets [Nagata et al, 1981b]. *Significantly different from the corresponding soy protein group.

with the baseline value, in the casein group than in the soy group. Since in the casein group there were no significant changes in concentrations in any of the lipoprotein fractions, the LDL cholesterol/apo-B ratio increased from 1.64 to 1.79. In contrast, due to a significant decrease in LDL cholesterol and increase in HDL cholesterol in soy protein group, this ratio declined from 1.90 to 1.79. It is thus suggested that ingesting the casein diet causes a shift to LDL particles richer in cholesterol, whereas the reverse is the case with the soy diet. This was partly confirmed by the changes in the density of LDL.

Roberts et al [1981] very recently reported effects of dietary proteins on the composition and turnover of apolipoproteins in plasma lipoproteins of rabbits. They prepared [^{125}I]-labeled intermediate-density lipoproteins (IDL) and LDL from rabbits fed semipurified diets containing either casein or soy protein, and reinjected it into rabbits fed one or the other of these two diets. Labeled apoproteins of IDL, isolated from rabbits on either diet, turned over more rapidly in rabbits fed soy protein compared with those fed casein. The most important observation is that labeled apoproteins of VLDL from soy protein-fed rabbits were transferred to HDL more rapidly than those from rabbits fed casein. This observation is of particular interest since it directly emphasizes that the dietary protein specifically modifies the structure and hence the metabolism of plasma lipoproteins. The concentration of apo-E was markedly increased in VLDL and IDL of casein-fed rabbits, and apo-C families also increased in VLDL. From these observations, Roberts et al [1981] suggested that the effects of dietary proteins on plasma cholesterol levels may be secondary to their effects on the concentration and metabolism of the protein components of plasma lipoproteins.

VIII. CONCLUDING REMARKS

The hypocholesterolemic or antihypercholesterolemic effect of soy protein compared with casein can in general be observed consistently in experimental animals such as rabbits and rats if the dietary regimens are appropriate. Nevertheless, several studies exist showing the ineffectiveness of dietary proteins in rats [Sautier et al, 1979; Neves et al, 1980; Pathirana et al, 1980]. The controversy is not surprising since the cholesterol-lowering action of plant protein varies depending on the type of proteins ingested and on the fat, carbohydrate, or other dietary components given at the same time. For example, studies with rats revealed that the antihypercholesterolemic effect of the soy protein isolate became less clear when the dietary corn oil level was increased from 1% to 5% in a cholesterol-free, semipurified diet [Nagata et al, 1980]. Therefore, the literature data should be evaluated with the greatest scrupulousness.

The mechanism of the hypocholesterolemic action in animal models appears to be nearly solved. The soy protein isolate as compared with casein causes the following changes in the cholesterol dynamics: (1) the reduction of intestinal absorption of cholesterol and/or bile acids and hence the increase in fecal excretion of steroids; (2) incomplete hepatic compensation for decreased delivery of cholesterol from the intestine by enhanced cholesterogenesis in the liver; (3) increase in the excretion of cholesterol and bile acids via the bile as a consequence of the above modulations, resulting in the stimulation of metabolic turnover of cholesterol in the body; (4) the possible reduction of cholesterol contents in the peripheral tissues including the artery; and (5) the depression of lipoprotein secretion by the liver and presumably by the intestine. Thus cholesterol dynamics appear to be preferably shifted in the direction of lowering the pool size of cholesterol not only in the plasma compartment but also in the tissues.

In contrast, when the amino acid mixture equivalent to soy protein was substituted for the intact protein, neither the decrease in the absorption of cholesterol and possibly bile acids (and hence the increase in steroid excretion) nor the reduction of hepatic secretion of cholesterol could be observed as compared with the casein-type amino acid mixture. The turnover rate of serum cholesterol also remained unchanged. In this case the decreased hepatic sterogenesis seems to be the most probable factor responsible for the reduced serum cholesterol level due to the soy protein-type mixture. The unchangeability of the sterol excretion in response to dietary amino acid mixtures is consistent with that observable in human studies. Thus, the studies with amino acid mixture diets in animal models may provide a clue to the underlying mechanism for plasma cholesterol-lowering action of soy protein in humans.

More dynamic as well as static analyses of the protein components of plasma lipoproteins are necessary for the elucidation of the precise mechanism involved in the hypocholesterolemic efficacy of soy protein, in particular in humans, since the available information suggests that dietary protein-dependent changes in plasma cholesterol levels may be secondary to the effects on the composition and metabolism of apolipoproteins [Roberts et al, 1981]. Since changes in the serum concentration of cholesterol produced by dietary proteins are the result of overall alteration in the balance of cholesterol dynamics, further insight into each parameter of cholesterol metabolism in humans is, of course, indispensable. It is at least conceivable that changes in hormonal status may serve as a trigger for the alteration of lipid metabolism and thereby the reduction of plasma lipid levels and the degree of atherogenesis.

Finally, the effects of dietary protein types on the lipid peroxidation should not have been disregarded. Casein, which contains a larger amount of easily oxidized amino acids (eg, histidine, Lys, Tyr) than soy protein, might increase the level of free-radical reactions [Harman, 1978]. A rise in the serum level of thiobarbituric acid-reactive substances on consuming casein compared with soy protein has been observed, though not consistently [Nagata et al, 1981a; Sugano et al, 1982a]. Available information is indicative of the causal relationship between lipid peroxidation and cardiovascular pathology. In this context, studies on effects of the source of dietary protein on serum tocopherol levels as performed by Eklund and Sjöblom [1980] are of interest.

ACKNOWLEDGMENTS

I wish to thank Dr Y. Nagata for his helpful suggestions during the preparation of this manuscript. This work was supported in part by Grant-in-Aid for Specific Project Research (412008) from the Ministry of Education, Culture, and Science and from the Research Committee on Soy Protein Nutrition.

IX. REFERENCES

Anderson JT, Grande F, Keys A (1971): Effect on man's serum lipids of two proteins with different amino acid composition. Am J Clin Nutr 24:524–530.

Assam R, Attali JR, Ballerio G, Boillot J, Girard AJ (1977): Glucagon secretion induced by natural and artificial amino acids in the perfused rat pancreas. Diabetes 26:300–307.

Aust L, Poledne R, Elhabet A, Noack R (1980): The hypolipaemic action of glycine rich diet in rats. Nahrung 24:663–671.

Bazzano G, D'Elia JA, Olson RE (1970): Monosodium glutamate: Feeding of large amounts in man and gerbils. Science 169:1208–1209.

Bosisio E, Galli-Kienle M, Galli G, Ghiselli GC, Franceschini G, Sirtori CR (1980): Experimental studies on the mechanism of the hypocholesterolemic effect of soy protein. VIIth International Symposium on Drugs Affecting Lipid Metabolism, May, Milan, Abstract Book, p 274.

Brown MS, Kovanen PT, Goldstein JL (1981): Regulation of plasma cholesterol by lipoprotein receptors. Science 212:628–635.

Calvert GD, Blight L, Illman RJ, Topping DL, Potter JD (1981): A trial of the effects of soya-bean flour and soya-bean saponins on plasma lipids, faecal bile acids and neutral sterols in hypercholesterolemic men. Br J Nutr 45:277–281.

Carroll KK (1981): Soy protein and atherosclerosis. J Am Oil Chem Soc 58:416–419.

Carroll KK, Giovannetti PM, Huff MW, Moase O, Roberts DCK, Wolfe BM (1978): Hypocholesterolemic effect of substituting soybean protein for animal protein in the diet of healthy young women. Am J Clin Nutr 31:1312–1321.

Carroll KK, Hamilton RMG (1975): Effects of dietary protein and carbohydrate on plasma cholesterol levels in relation to atherosclerosis. J Food Sci 40:18–23.

Carroll KK, Huff MW, Roberts DCK (1977): Dietary protein, hypercholesterolemia, and atherosclerosis. In Schettler G, Goto Y, Hata Y, Klose G (eds): "Atherosclerosis IV." Berlin, Heidelberg, New York: Springer-Verlag, pp 445–448.

Cittadini D, Pietropalo C, DeCristofaro D, D'Ayjello-Caracciole M (1964): *In vitro* effect of L-lysine on rat liver arginase. Nature 203:643–644.

Connor WE, Connor SL (1972): The key role of nutritional factors in the prevention of coronary heart disease. Prev Med 1:49–83.

Czarnecki S, Kritchevsky D (1980): The effect of dietary proteins on lipoprotein metabolism. J Am Oil Chem Soc 56:388A.

Drane HM, Patterson DSP, Roberts BA, Saba N (1980): Oestrogenic activity of soya-bean products. Food Cosmet Toxicol 18:425–427.

Dugan RE, Ness GC, Lakshmann MR, Nepokroeff CM, Porter JW (1974): Regulation of hepatic β-hydroxy-β-methylglutaryl coenzyme A reductase by the interplay of hormones. Arch Biochem Biophys 161:499–504.

Eaton R (1973): Hypolipemic action of glucagon in experimental endogenous lipidemia in the rat. J Lipid Res 14:312–318.

Eisenstein AB, Strack I, Gallo-Torres H, Georgiadis A, Miller ON (1979): Increased glucagon secretion in protein-fed rats: Lack of relationship to plasma amino acids. Am J Physiol 236: E20–E27.

Eklund A, Sjöblom L (1980): Effects of the source of dietary protein on serum low density lipoproteins (VLDL + LDL) and tocopherol levels in female rats. J Nutr 110:2321–2335.

Fajans SS, Floyd JC Jr, Knopf RF, Conn JW (1968): Effect of amino acids and proteins on insulin secretion in man. Recent Prog Hor Res 23:617–662.

Forsythe WA, Miller ER, Hill GM, Romsos DR, Simpson RC (1980): Effect of dietary protein and fat sources on plasma cholesterol parameters, LCAT activity and amino acid levels and on tissue lipid content of growing pigs. J Nutr 110:2467–2479.

Fumagalli R, Paoletti R, Howard AN (1978): Hypocholesterolaemic effect of soya. Life Sci 22: 947–952.

Garlich JD, Bazzano G, Olson RE (1970): Changes in plasma free amino acid concentrations in human subjects on hypocholesterolemic diets. Am J Clin Nutr 23:1626–1638.

Glickman RM (1980): Intestinal lipoprotein formation. Nutr Metab 24 (suppl 1):3–11.

Goodman DS, Noble RP (1968): Turnover of plasma cholesterol in man. J Clin Invest 47:231–241.

Goulding NJ, Gibney MJ, Gallagher PJ, Jones DB, Morgan JB, Taylor TG (1980): The immunological effect of a high dietary intake of soya protein in man. Proc Nutr Soc 39:2A.

Grundy SM (1975): Effect of polyunsaturated fats on lipid metabolism in patients with hyperglyceridemia. J Clin Invest 55:269–282.

Grundy SM, Abrams JJ (1981): Comparison of soy protein and casein on plasma lipids and lipoproteins in man. In: "Symposium on Soy Protein and Polysaccharide in Human Nutrition and Health." Keystone, Colorado, August.

Harman D (1978): Free radical theory of aging: Nutritional implications. Age 1:145–152.

Havel RJ (1980): Lipoprotein biosynthesis and metabolism. Ann NY Acad Sci 348:16–29.

Hermus RJJ, Dallinga-Thie GM (1979): Soya, saponins, and plasma-cholesterol. Lancet ii:48.

Hevia P, Kari FW, Ulman EA, Visek WJ (1980a): Serum and liver lipids in growing rats fed casein with L-lysine. J Nutr 110:1224–1230.

Hevia P, Ulman EA, Kari FW, Visek WJ (1980b): Serum lipids of rats fed excess L-lysine and different carbohydrates. J Nutr 110:1231–1239.

Holmes WL, Rubel GB, Hood SS (1980): Comparison of the effect of dietary meat versus dietary soybean protein on plasma lipids of hyperlipidemic individuals. Atherosclerosis 36:379–387.

Huff MW, Carroll KK (1980a): Effects of dietary protein on turnover, oxidation, and absorption of cholesterol, and on steroid excretion in rabbits. J Lipid Res 21:546–558.

Huff MW, Carroll KK (1980b): Effects of dietary proteins and amino acid mixtures on plasma cholesterol levels in rabbits. J Nutr 110:1676–1685.

Huff MW, Hamilton RMG, Carroll KK (1977): Plasma cholesterol levels in rabbits fed low fat, cholesterol-free semipurified diets: Effects of dietary proteins, protein hydrolysates and amino acid mixtures. Atherosclerosis 28:187–195.

Ignatowski A (1909): Uber die Wirkung des tierischen Eiweisses auf die Aorta und die parenchymatösen Organe der Kaninchen. Virchow Arch Pathol Anat Physiol Klin Med 198:248–270.

Jarowski CI, Pytelewski R (1975): Utility of fasting essential amino acid plasma levels in formulation of nutritionally adequate diets. III Lowering of rat serum cholesterol levels by lysine supplementation. J Pharm Sci 64:690–691.

Katan MB, Vroomen L, Hermus RJJ (1982): Reduction of casein-induced hypercholesterolemia and atherosclerosis in rabbits and rats by dietary glycine, arginine and alanine. Atherosclerosis 43:381–391.

Kim DN, Lee KT, Reiner JM, Thomas WA (1978): Effects of a soy product on serum and tissue cholesterol concentrations in swine fed high-fat, high-cholesterol diets. Exp Mol Pathol 29:385–399.

Kim DN, Lee KT, Reiner JM, Thomas WA (1980a): Increased steroid excretion in swine fed high-fat, high-cholesterol diet with soy protein. Exp Mol Pathol 33:25–35.

Kim DN, Rogers DH, Li JR, Lee KT, Reiner JM, Thomas WA (1980b): Some effects of a grain-based mash diet on cholesterol metabolism in swine. Exp Mol Pathol 32:143–153.

Kimura S, Suwa J, Ito M, Sato H (1979): Experimental studies on the role of defatted soybean in the development of malignant goiter. In Miller EC (ed): "Naturally Occurring Carcinogens-Mutagens and Modulators of Carcinogenesis." Tokyo: Japan Scientific Society Press/Baltimore: University Park Press, pp 101–110.

Kritchevsky D (1979): Vegetable protein and atherosclerosis. J Am Oil Chem Soc 56:135–140.

Kritchevsky D, Klurfeld DM (1979): Influence of vegetable protein on gallstone formation in hamsters. Am J Clin Nutr 32:2174–2176.

Kritchevsky D, Tepper SA, Czarnecki S, Klurfeld DM, Story JA (1981): Experimental atherosclerosis in rabbits fed cholesterol-free diets. Part 9. Beef protein and textured vegetable protein. Atherosclerosis 39:169–175.

Lacomb C, Nibbelink M (1980): Lipoprotein modification with changing dietary proteins in rabbits on a high fat diet. Artery 6:280–289.

Leiner IE (1977): Nutritional aspects of soy protein products. J Am Oil Chem Soc 54:454A–472A.

Liepa GU, Park M (1981): Role of oilseed protein in lipoprotein metabolism. J Am Oil Chem Soc 58:608A.

McGuinness EE, Morgan RGH, Levison DA, Frape DL, Hopwood D, Wormsley KG (1980): The effects of long-term feeding of soya flour on the rat pancreas. Scand J Gastroenterol 15: 497–502.

Melish J, Le NA, Ginsberg H, Steinberg D, Brown WV (1980): Dissociation of apoprotein B and triglyceride production in very-low-density lipoproteins. Am J Physiol 239:E354–362.

Mokady S, Einav P (1978): Effect of dietary wheat gluten on lipid metabolism in growing rats. Nutr Metab 22:181–189.

Munoz JM, Sandstead HH, Jacob RA, Logan GM, Reck SJ, Klevay LM, Dintzis FR, Inglett GE, Shuey WC (1979): Effects of some cereal brans and textured vegetable protein on plasma lipids. Am J Clin Nutr 32:580–592.

Nagata Y, Imaizumi K, Sugano M (1980): Effects of soya-bean protein and casein on serum cholesterol levels in rats. Br J Nutr 44:113–121.

Nagata Y, Ishiwaki N, Sugano M (1982): Studies on the mechanism of antihypercholesterolemic action of soy protein and soy protein-type amino acid mixture in relation to the casein counterparts in rats. J Nutr 112:1614–1625.

Nagata Y, Tanaka K, Sugano M (1981a): Further studies on the hypocholesterolaemic effect of soya-bean protein in rats. Br J Nutr 45:233–241.

Nagata Y, Tanaka K, Sugano M (1981b): Serum and liver cholesterol levels of rats and mice fed soy-bean protein or casein. J Nutr Sci Vitaminol (Tokyo) 27:583–593.

Nestel PJ, Havenstein N, Homma Y, Scott TW, Cook LJ (1975): Increased sterol excretion with polyunsaturated-fat high-cholesterol diets. Metabolism 24:189–198.

Nestel PJ, Whyte HM, Goodman DS (1969): Distribution and turnover of cholesterol in humans. J Clin Invest 48:982–991.

Neves LB, Clifford CK, Kohler GO, DeFremery D, Knuckles BE, Cheowtirakul C, Miller MW, Weir WC, Clifford AJ (1980): Effects of dietary proteins from a variety of sources on plasma lipids and lipoproteins of rats. J Nutr 110:732–742.

Nilsson-Ehle P (1981): Lipolytic enzymes and plasma lipoprotein metabolism. Annu Rev Biochem 49:667–693.

Noseda G, Fragiacomo C (1980): Effects of soybean protein diet on serum lipids, plasma glucagon, and insulin. In Noseda G, Levis B, Paoletti R (eds): "Diet and Drugs in Atherosclerosis." New York: Raven Press, pp 61–65.

Noseda G, Fragiacomo C, Descovich GC, Fumagalli R, Bernini F, Sirtori CR (1980): Clinical studies on the mechanism of action of the soybean protein diet. In Fumagalli R, Kritchevsky D, Paoletti R (eds): "Drugs Affecting Lipid Metabolism." Amsterdam, New York, Oxford: Elsevier/North-Holland, pp 355–362.

Okita T, Sugano M (1981): Effects of dietary soybean globulins on plasma and liver lipids and on fecal excretion of neutral sterols in rats. J Nutr Sci Vitaminol (Tokyo) 27:379–388.

Olson RE, Nichaman MZ, Nittka J, Eagles JA (1970): Hypocholesterolemic effects of amino acid diets in man. Am J Clin Nutr 23:1614–1625.

Pathirana C, Gibney MJ, Taylor TG (1980): Effects of soy protein and saponin on serum and liver cholesterol in rats. Atherosclerosis 36:595–596.

Potter JD, Illman RJ, Calvert GD, Oakenfull DG, Topping DL (1980): Soya saponins, plasma lipids, lipoproteins and fecal bile acids: A double blind cross-over study. Nutr Rep Int 22: 521–528.

Potter JM, Med B, Nestel PJ (1976): Greater bile acid excretion with soy bean than with cow milk in infants. Am J Clin Nutr 29:546–551.

Raaij JMA van, Katan MB, Hautvast JGAJ, Hermus RJJ (1981): Effects of casein versus soy protein diets on serum cholesterol and lipoproteins in young healthy volunteers. Am J Clin Nutr 34:1261–1271.

Raheja KL, Linscheer WG (1980): Hypolipidemic effect of casein vs. soy protein in the hyperlipidemic hypothyroid chick model. Nutr Rep Int 24:497–503.

Raja K, Jarowski CI (1975): Utility of fasting essential amino acid plasma levels in formulation of nutritionally adequate diets. IV Lowering of human plasma cholesterol and triglyceride levels by lysine and tryptophan supplementation. J Pharm Sci 64:691–692.

Reaven GM, Lerner RL, Stern MP, Farquhar JW (1967): Role of insulin in endogenous hypertriglyceridemia. J Clin Invest 46:1756–1767.

Reiser R, Henderson GK, O'Brien BC, Thomas J (1977): Hepatic 3-hydroxy-3-methyl-glutaryl coenzyme-A reductase of rats fed semipurified and stock diets. J Nutr 107:453–457.

Richter W, Weisweiler P, Schwandt P (1980): The effect of a mixture of soy bean protein and pectin. VIIth International Symposium on Drugs Affecting Lipid Metabolism, May, Milan, Abstract Book, p 101.

Roberts DCK, Huff MW, Carroll KK (1979): Influence of diet on plasma cholesterol concentrations in suckling and weanling rabbits. Nutr Metab 23:476–486.

Roberts DCK, Stalmach ME, Khalil WM, Hutchinson JC, Carroll KK (1981): Effects of dietary protein on composition and turnover of apoproteins in plasma lipoproteins of rabbits. Can J Biochem 59:642–647.

Roy DM, Schneeman BO (1981): Effect of soy protein, casein and trypsin inhibitor on cholesterol, bile acids and pancreatic enzymes in mice. J Nutr 111:878–885.

Sautier C, Doucet C, Flament C, Lemonnier D (1979): Effects of soy protein and saponins on serum, tissue and feces steroids in rat. Atherosclerosis 34:233–241.

Schade DS, Woodside W, Eaton RP (1979): The role of glucagon in the regulation of plasma lipids. Metabolism 28:874–886.

Schaefer EJ, Eisenberg S, Levy RI (1978): Lipoprotein apoprotein metabolism. J Lipid Res 19:667–687.

Shore B, Shore V, Salel A, Mason D, Zelis R (1974a): An apolipoprotein preferentially enriched in cholesteryl ester-rich very low density lipoproteins. Biochem Biophys Res Commun 58:1–7.

Shore VG, Shore B, Hart RG (1974b): Changes in apolipoproteins and properties of rabbit very low density lipoproteins on induction of cholesteremia. Biochemistry 13:1579–1584.

Shorey RL, Bagan B, Lo GS, Steinke FH (1981): Determinants of hypocholesterolemic response to soy and animal protein based-diets. Am J Clin Nutr 34:1769–1778.

Silk DBA (1980): Digestion and absorptiion of dietary protein in man. Proc Nutr Soc 39:61–70.

Sirtori CR, Descovich G, Noseda G (1980): Textured soy protein and serum-cholesterol. Lancet i:149.

Sleisenger MH, Kim YS (1979): Protein digestion and absorption. New Engl J Med 300:659–663.

Smith LC, Pownall HJ, Gotto AM Jr (1978): The plasma lipoproteins: Structure and metabolism. Annu Rev Biochem 47:751–777.

Srikant CB, McCorkle K, Unger RH (1977): Properties of immunoreactive glucagon fractions of canine stomach and pancreas. J Biol Chem 252:1847–1851.

Sugano M, Ishiwaki N, Nagata Y, Imaizumi K (1982a): Effects of arginine and lysine addition to casein and soya-bean protein on serum lipids, apolipoproteins, insulin and glucagon in rats. Br J Nutr 48:211–221.

Sugano M, Nagata Y, Tanaka K, Imaizumi K (1981): Effects of dietary proteins on the metabolism of lipoproteins in rats. XIIth International Congress of Nutrition, August, San Diego, Abstracts, p 128.

Sugano M, Tanaka K, Ide T (1982b): Secretion of cholesterol, triglyceride and apolipoprotein A-I by isolated perfused liver from rats fed soybean protein and casein or their amino acid mixtures. J Nutr 112:855–862.

Terpstra AHM, Herms RJJ, West CE (1982): The role of dietary protein in cholesterol metabolism. World Rev Nutr Diet (in press).

Terpstra AHM, Sanchez-Muniz FJ (1981): Time course of the development of hypercholesterolemia in rabbits fed semipurified diets containing casein or soybean protein. Atherosclerosis 39:217–227.

Torre GM, Lynch VP, Jarowski CI (1980): Lowering of serum cholesterol and triglyceride levels by balancing amino acid intake in the white rat. J Nutr 110:1194–1196.

Waggle DH, Kolar CW (1979): Types of soy protein products. In Wilcke HL, Hopkins DT, Waggle DH (eds): "Soy Protein and Human Nutrition." New York: Academic Press, pp 19–51.

Weigensberg BI, Stary HC, McMillan GC (1964): Effect of lysine deficiency on cholesterol atherosclerosis in rabbits. Exp Mol Pathol 3:444–454.

Yadav NR, Liener IE (1977): Reduction of serum cholesterol in rats fed vegetable protein or an equivalent amino acid mixture. Nutr Rep Int 16:385–389.

Yudkin J (1957): Diet and coronary thrombosis. Hypothesis and fact. Lancet ii:155–162.

Animal and Vegetable Proteins in Lipid
Metabolism and Atherosclerosis, pages 85–100
© 1983 Alan R. Liss, Inc., 150 Fifth Ave., New York, NY 10011

5
Effects of Animal and Vegetable Protein in Experimental Atherosclerosis

David Kritchevsky, Shirley A. Tepper, Susanne K. Czarnecki,
David M. Klurfeld, and Jon A. Story
The Wistar Institute of Anatomy and Biology, 36th Street at Spruce, Philadelphia,
Pennsylvania 19104

The first comparison of the atherogenic effects of animal and vegetable protein was carried out by Meeker and Kesten [1940, 1941]. They maintained rabbits on a basal diet (wheat flour, alfalfa-leaf meal, linseed meal, salt mix, and ground carrots) which provided (by weight) 15% protein, 55% carbohydrate, and 5% fat. They compared this diet with one high (38%) in animal protein (casein, wheat flour, alfalfa-leaf and linseed meals, salt mix, and carrots) and one high (39%) in vegetable protein. The vegetable protein contained soy flour, the other ingredients being the same as those present in the basal and high-animal-protein diets. The percentage of calories provided by protein, carbohydrate, and fat in the three diets was as follows: basal–protein, 18.3%, carbohydrate, 67.1%, and fat, 14.6%; casein–protein, 44.6%, carbohydrate, 45.6%, and fat, 9.8%; and soy protein–protein, 49.3%, carbohydrate, 42.6%, and fat, 8.1%. Some of the animals received a daily dose of 60 or 250 mg of cholesterol in 1 ml of olive oil. The cholesterol supplement was fed for three months and a cholesterol-free diet for a further three months. Their findings are summarized in Table 5-I. It is evident that soy protein was significantly less atherogenic than casein. Few plasma cholesterol data were given

Susanne Czarnecki's present address is National Heart, Lung and Blood Institute, National Institutes of Health, Bethesda, MD 20205.

Jon Story's present address is Department of Foods and Nutrition, Purdue University, West Lafayette, IN 47907.

TABLE 5-I. Influence of Casein or Soy Protein on Atherosclerosis in Rabbits

Regimen	Cholesterol (mg)	Number of rabbits	Percent sclerotic	Average severity[a]
Basal	—	8	100	0
Casein	—	12	50	0.75
Soy	—	8	0	0
Basal	60	21	70	1.24
Casein	60	13	75	2.08
Soy	60	16	35	0.44
Basal	60	9	56	0.89
Casein	60	13	77	2.08
Soy	60	6	33	0.33
Basal	250	6	67	1.50
Soy	250	6	50	0.67
Basal[b]	250	6	100	1.50
Soy	250	4	25	0.25

[a]Graded on 0–3 scale.
[b]Three-month experiment, others six months.
After Meeker and Kesten [1940, 1941].

but extrapolation from a bar graph suggests that plasma cholesterol levels (mg/dl) in rabbits fed the cholesterol-free diets were basal, 53 ± 11; casein, 125 ± 13; and soy protein, 64 ± 9.

Enselme et al [1963] compared casein and wheat gluten in a diet which contained 25% protein and 8% corn oil. Casein was significantly more cholesterolemic, 120 ± 7 mg/dl vs 77 ± 8 mg/dl. There were no aortic lesions, but the aortic free/esterified cholesterol ratio was 0.36 in the wheat gluten group and 0.16 in the casein group. The ratio of aortic free/esterified cholesterol falls with increasing severity of atherosclerosis [Newman and Zilversmit, 1964; Kritchevsky, 1967].

Howard et al [1965] fed rabbits a semipurified diet containing 25% casein (24.5% of calories) and 20% beef fat (44.1% of calories). Their results are summarized in Table 5-II. Casein was significantly more cholesterolemic and atherogenic than soy protein. Replacement of casein with soy flour, which may have contained less protein and more carbohydrate, virtually eliminated atherosclerosis (present in only one of 11 rabbits).

Our current interest in protein effects was stimulated by our investigations into the interactions among dietary constituents as they relate to induction of atherosclerosis. Lofland et al [1966] investigated the effect of protein on atherosclerosis and cholesterolemia in white Carneau pigeons. They compared the effects of wheat gluten and casein-lactalbumin 85:15 in birds fed butter, corn oil, shortening, or margarine. The effects were not predictable (Table

TABLE 5-II. Influence of Animal or Vegetable Protein on Atherosclerosis in Rabbits

Regimen	Number of rabbits	Mean plasma cholesterol (mg/dl)	Average atherosclerosis
Casein	10	391	2.30 ± 0.21
Soy protein	4	292	1.00 ± 0.41
Soy flour	11	127	0.09 ± 0.09
Control	12	103	0

After Howard et al [1965].

TABLE 5-III. Influence of Protein (8% of Calories) and Fat (30% of Calories) on Cholesterolemia and Atherosclerosis in White Carneau Pigeons

	Fat[a]			
	Butter	Corn oil	Shortening	Margarine
Serum cholesterol (mg/dl)				
Casein-lactalbumin[b]	472	345	577	375
Wheat gluten	419	492	495	442
Percent of birds with lesions				
Casein-lactalbumin	89	100	89	73
Wheat gluten	100	78	85	60
Atherosclerotic index				
Casein-lactaibumin	5.1	5.2	3.5	3.0
Wheat gluten	3.5	5.7	4.3	2.0

[a]Diets contained 30 mg cholesterol/100 g fat.
[b]Ratio 85:15.
After Lofland et al [1966].

5-III). When the fat was more saturated (butter, shortening) birds fed animal protein had higher cholesterol levels but not higher incidence of atherosclerosis. Animal protein led to a greater incidence of lesions in birds fed the more unsaturated fats (corn oil, margarine).

We examined the influence of different types of dietary fiber on the atherogenic effects of animal or vegetable protein. In this experiment, as in all others to be discussed, we used a semipurified, cholesterol-free diet which is hypercholesterolemic and atherogenic for rabbits [Kritchevsky and Tepper, 1965, 1968]. This diet leads to endogenous hyperlipidemia and vitiates the justifiable criticism of feeding cholesterol to herbivores as a means of establishing atherosclerosis. The diet contains 40% carbohydrate, 25% protein, 15% fiber, 14% fat, 5% salt mix, and 1% vitamin mix. The caloric distribution (% of calories) is carbohydrate, 41.5%; fat, 32.6%; and protein, 25.9%. The standard ingredients are sucrose, casein, cellulose, and coconut oil.

In the study of fiber-protein interaction [Kritchevsky et al, 1977] we compared cellulose, wheat straw, and alfalfa in rabbits fed either casein or soy protein. Wheat straw had been shown to inhibit atherosclerosis in rabbits fed a semipurified diet containing butter [Moore, 1967]. Cookson et al [1967] and Cookson and Fedoroff [1968] had described the hypocholesterolemic effects of alfalfa. Our results are summarized in Table 5-IV. When the dietary fiber was cellulose, casein was considerably more cholesterolemic and atherogenic than soy protein. Substitution of wheat straw for cellulose did not influence differences in lipidemia and reduced the difference in atherogenicity by a slight degree. When the fiber was alfalfa there were no differences in cholesterolemia or degree of atherosclerosis.

Interaction of protein with dietary components other than fiber can also affect cholesterolemia and atherosclerosis. We [Kritchevsky et al 1976] compared the effects of a semipurified diet containing casein, sucrose, coconut oil, and cellulose with one containing skim milk powder, coconut oil, and cellulose. The two diets had similar levels of protein and carbohydrate. One principal difference between the diets was that the carbohydrate in the skim milk diet was mainly lactose. Serum cholesterol levels (mg/dl ± SEM) were 402 ± 40 in the casein group and 337 ± 44 in the rabbits fed skim milk powder. Average atherosclerosis (arch plus thoracic/2) ± SEM was casein 1.5 ± 0.27 and skim milk powder 0.75 ± 0.12 ($P < 0.05$).

We also investigated the effects of other plant and animal proteins on atherosclerosis. In one case (Table 5-V) we compared the effects of corn protein, wheat gluten, and lactalbumin [Czarnecki, 1982; Kritchevsky, 1979; Kritchevsky et al, 1982b]. The animal protein was twice as cholesterolemic and was 83–175% more atherogenic than the vegetable proteins. The serum lipoproteins were separated on an agarose column [Rudel et al, 1974] with the results shown in Table 5-VI. Rabbits fed lactalbumin had about twice the concentration of very low density (VLDL) and intermediate-density (IDL) lipoproteins;

TABLE 5-IV. Influence of Dietary Fiber on Cholesterolemia and Atherosclerosis in Rabbits Fed Casein or Soy Protein*

| Protein (25%) | Fiber (15%) | Serum lipids (mg/dl ± SEM) | | Average atherosclerosis | |
		Cholesterol	Triglycerides	Arch	Thoracic
Casein	Cellulose	402 ± 40[a,b]	164 ± 45[f,g]	1.81 ± 0.2[h]	1.19 ± 0.23
Soy	Cellulose	248 ± 44[a]	41 ± 8[f]	1.50 ± 0.39	1.00 ± 0.52
Casein	Wheat straw	375 ± 42[c,d]	90 ± 19	1.17 ± 0.22	0.88 ± 0.18
Soy	Wheat straw	254 ± 35[c,e]	66 ± 9	1.04 ± 0.28	0.77 ± 0.24
Casein	Alfalfa	193 ± 34[b,d]	60 ± 8[g]	0.70 ± 0.11[h]	0.55 ± 0.20
Soy	Alfalfa	159 ± 20[e]	62 ± 17	0.88 ± 0.22	0.58 ± 0.17

*Diets also contained 40% sucrose and 14% coconut oil. Fed 10 months.
[a-h]Values bearing the same letter are significantly different ($P < 0.05$).
After Kritchevsky et al [1977].

TABLE 5-V. Influence of Corn Protein, Wheat Gluten and Lactalbumin
on Atherosclerosis in Rabbits

	Protein[a]		
	Corn protein	Wheat gluten	Lactalbumin
Number	7	8	7
Serum lipids (mg/dl ± SEM)			
Cholesterol	158 ± 25	152 ± 18	312 ± 79
Triglycerides	94 ± 9	98 ± 13	122 ± 52
Phospholipids	84 ± 3	91 ± 3	84 ± 3
Liver lipids (g/100 g ± SEM)			
Cholesterol	1.08 ± 0.18	1.11 ± 0.14	1.05 ± 0.12
Triglycerides	2.43 ± 0.64	3.18 ± 0.94	1.52 ± 0.40
Phospholipids	1.83 ± 0.04	1.97 ± 0.04	1.91 ± 0.06
Atherosclerosis (± SEM)			
Arch	0.4 ± 0.17	0.6 ± 0.18	1.1 ± 0.37
Thoracic	0.2 ± 0.15	0.4 ± 0.08	0.6 ± 0.18

[a]Diet contained 40% sucrose, 25% protein, 15% cellulose, 14% coconut oil. Fed 8 months.
After Kritchevsky et al [1982a].

TABLE 5-VI. Distribution of Lipoproteins in Sera of Rabbits Fed Various Proteins

	Lipoprotein class (μg/ml [%])			
Diet	VLDL	IDL	LDL	HDL
Corn protein	26 (3.3)	130 (16.3)	178 (22.4)	462 (58.0)
Wheat gluten	10 (2.5)	41 (10.2)	145 (36.1)	206 (51.2)
Lactalbumin	84 (5.8)	382 (26.4)	333 (23.0)	648 (44.8)

After Czarnecki [1982] and Kritchevsky et al [1982a].

sera of rabbits fed wheat gluten carried a larger proportion of low-density lipoprotein (LDL); the concentration of high-density lipoprotein (HDL) was
lowest in sera of rabbits fed lactalbumin.

In seeking a possible explanation for the different effects of animal and
vegetable protein Huff et al [1977] fed mixtures of the L-amino acids compris-
ing casein and soy protein. They found no effect of the casein mixture and an
80% increase in serum cholesterol in rabbits fed the soy amino acid mixture.
Feeding enzymic (partial) hydrolyzates of casein and soy protein gave lower
serum cholesterol levels, by 16% and 41%, respectively. Feeding various
amino acid mixtures [Huff and Carroll, 1980a] showed that a few (casein essential amino acids plus glutamic acid; casein essential amino acids plus serine,
aspartic acid, proline, and glutamic acid) reduced cholesterol levels below
those seen in casein feeding. No amino acid combination was less cholesterolemic than soy isolate and a few (soy essential amino acids plus alanine or glu-

tamic acid) were significantly more cholesterolemic. It was not possible to localize the effects to any specific amino acid. We examined the possibility that the lysine/arginine ratio might play a role. Our line of reasoning was that since lysine is known to antagonize arginine metabolism in rats [Jones et al, 1966] and chicks [Jones, 1964], it might also do so in the rabbit. Lysine inhibits arginase activity [Hunter and Downs, 1945; Cittadini et al, 1964] and arginine deficiency leads to fatty liver formation [Milner et al, 1975; Milner and Perkins, 1978; Milner and Hassan, 1981]. It is possible then that with more lysine in the diet, less arginine will be metabolized and more will be available for incorporation into the arginine-rich lipoprotein which is atherogenic for rabbits [Shore et al, 1974]. Weigensberg et al [1964] had suggested that the lysine content of a protein could be a major determinant of its atherogenicity. They fed rabbits diets containing 1 g of cholesterol, 6 g of elaidinized olive oil, and either 552, 52, or 22 mg/day of lysine. The lysine/arginine ratios of the diets used were 2.16, 0.20, and 0.09, respectively. The lysine-replete diet was the more atherogenic, but the other two diets were severely deficient in this amino acid.

In order to test this hypothesis we carried out a series of three experiments in which we compared the atherogenic effects of casein (CAS), soy protein (SOY), casein plus arginine sufficient to give the lysine/arginine ratio of soy protein (CAS-A) and soy protein plus enough lysine to give the lysine/arginine ratio of casein (SOY-L). Various aspects of these studies have been reported [Czarnecki, 1982; Czarnecki and Kritchevsky, 1979, 1980; Kritchevsky, 1979, 1980; Kritchevsky et al, 1978]. Data from the individual experiments are tabulated in Table 5-VII. When the three experiments are combined we find that addition of arginine to casein did not affect cholesterolemia or triglyceridemia but reduced the severity of aortic atherosclerosis by 20% and of thoracic atherosclerosis by 30%. Addition of lysine to soy protein did not affect serum triglyceride levels but increased cholesterol levels by 53%; the severity of aortic atherosclerosis was increased by 57% and of thoracic atherosclerosis by 75%.

The serum lipoproteins were separated on an agarose column. The total serum lipoproteins (μg/ml) in the four groups of rabbits were CAS, 904; SOY, 807; CAS-A, 1,130; and SOY-L, 672. The percentage distribution of the various lipoprotein classes is shown in Table 5-VIII. The percentages of VLDL and IDL present in sera of rabbits fed CAS or SOY-L were similar and double those observed in the other two groups. The percentages of LDL were similar and those of HDL lowest in groups CAS and SOY-L.

We also compared the effects of three proteins (fish, casein, and whole milk), which had similar lysine content but differed in the amount of arginine present. The lysine present in the three proteins was fish, 6.81; casein, 6.91; and whole milk, 6.61. The lysine/arginine ratios (L/A) were: fish, 1.44; casein, 1.89; and whole milk, 2.44. The results of an 8 month trial are given in

TABLE 5-VII. Influence of Lysine and Arginine Added to Soy Protein and Casein on Experimental Atherosclerosis in Rabbits (Data ± SEM)*

Protein	Number of rabbits	Serum lipids (mg/dl)		Average atherosclerosis	
		Cholesterol	Triglyceride	Arch	Thoracic
Experiment 1					
CAS	7	174 ± 30[a]	133 ± 17	2.2 ± 0.5	1.5 ± 0.4
SOY	7	77 ± 21[a]	98 ± 17[a]	1.1 ± 0.4	0.7 ± 0.3
CAS-A	6	129 ± 12	186 ± 20[a,b]	1.4 ± 0.4	0.8 ± 0.3
SOY-L	6	106 ± 29	101 ± 14[a]	1.6 ± 0.4	1.1 ± 0.2
Experiment 2					
CAS	8	283 ± 28	81 ± 11[c]	1.1 ± 0.3	0.8 ± 0.3
SOY	11	234 ± 20	53 ± 7[a]	0.5 ± 0.2	0.4 ± 0.1
CAS-A	7	343 ± 65	117 ± 58	1.3 ± 0.4	0.7 ± 0.2
SOY-L	11	242 ± 22	70 ± 7	0.9 ± 0.3	0.5 ± 0.2
Experiment 3					
CAS	5	377 ± 59[b]	104 ± 28	1.6 ± 0.2[a,b]	0.9 ± 0.3[a]
SOY	7	117 ± 10[b,c]	43 ± 4[d,e]	0.6 ± 0.2[a]	0.1 ± 0.1[a,b,c]
CAS-A	7	271 ± 74	75 ± 13[d]	1.2 ± 0.2	1.3 ± 0.2[b,d,]
SOY-L	8	242 ± 43[c]	60 ± 5[e]	0.5 ± 0.2[b]	0.7 ± 0.1[c,d]

*Diets contained 40% sucrose, 25% protein, 15% cellulose, 14% coconut oil. Lysine/arginine (L/A) ratio: CAS, 2.0; SOY, 0.9; CAS-A, 0.9; SOY-L, 2.0.
[a-e]Values in any vertical column bearing the same letter are significantly different ($P < 0.05$).

TABLE 5-VIII. Distribution of Lipoproteins in Sera of Rabbits Fed Various Proteins

Protein	Lipoprotein class (μg/ml [%])			
	VLDL	IDL	LDL	HDL
CAS	21 (2.3)	130 (14.4)	288 (31.9)	465 (51.4)
SOY	9 (1.1)	62 (7.7)	242 (30.0)	494 (61.2)
CAS-A	10 (0.9)	88 (7.8)	405 (35.8)	627 (55.5)
SOY-L	14 (2.2)	114 (17.8)	194 (30.2)	320 (49.8)

Table 5-IX [Kritchevsky et al, 1982a]. It is evident that fish protein was less lipidemic and atherogenic than the other two proteins. When the average atherosclerosis was plotted against L/A, it yielded a straight line ($r = 0.9979$, $P < 0.05$). The foregoing data demonstrate that specific amino acids content may play a role in determining cholesterolemic and atherogenic effects of that protein.

Carroll and Hamilton [1975] had demonstrated that within the classification of animal and vegetable proteins there was a wide variation in influence on cholesterolemia. Thus a diet containing 30% casein, fed to rabbits for 30 days, resulted in plasma cholesterol levels double those seen in rabbits fed 30% raw egg white for the same period. Similarly, a 30% wheat gluten diet was over

TABLE 5-IX. Influence of Fish Protein, Casein and Whole Milk Protein on Lipid Levels and Atherosclerosis in Rabbits*

	Dietary protein			ANOV
	Fish	Casein	Whole milk	$P <$
Number	10/12	10/12	9/12	NS[c]
Weight change (g)	−372 ± 94	−373 ± 117	−308 ± 69	NS
Liver weight (g)	35.8 ± 1.3	40.2 ± 3.6	41.8 ± 2.9	NS
Liver as % body weight	2.06 ± 0.03	2.36 ± 0.19	2.26 ± 0.10	NS
Serum (mg/dl)				
Cholesterol	283 ± 40[a,b]	530 ± 76[a]	462 ± 62[b]	0.05
% Esterified	67.3 ± 2.1	68.3 ± 2.4	65.8 ± 2.9	0.05
% HDL cholesterol	15.7 ± 1.4[c]	11.8 ± 1.1[c]	11.9 ± 1.5	0.001
Triglyceride	122 ± 20[d]	177 ± 47	251 ± 56[d]	0.001
Phospholipid	134 ± 24[e,f]	238 ± 35[e]	249 ± 29[f]	0.05
Protein (g/dl)	4.00 ± 0.14	4.05 ± 0.18	4.14 ± 0.21	NS
Liver (g/100 g)				
Cholesterol	0.75 ± 0.05	0.90 ± 0.07	1.06 ± 0.14	NS
% Esterified	54.7 ± 1.1	59.4 ± 2.5	50.9 ± 3.4	0.001
Triglyceride	2.41 ± 0.26	1.81 ± 0.16	2.34 ± 0.36	NS
Phospholipid	1.43 ± 0.07	1.28 ± 0.07[g]	1.60 ± 0.12[g]	0.025
Protein	19.70 ± 1.04[h]	14.66 ± 0.54[h,i]	17.22 ± 0.92[i]	0.001
Atherosclerosis				
Arch	1.55 ± 0.23[j]	2.05 ± 0.25	2.61 ± 0.16[j]	NS
Thoracic	0.95 ± 0.17[k]	1.10 ± 0.25	1.56 ± 0.19[k]	NS

*Rabbits fed semipurified diets containing 25% protein for eight months. All data ± SEM. ANOV, analysis of variance; NS, not significant.

[a-k]Values bearing the same letter are significantly ($P \leq 0.05$) different by t-test.

five times as cholesterolemic as one containing soy protein isolate. We compared the effects of beef protein and casein on atherogenesis in rabbits. The beef was dehydrated but not defatted, thus we used beef tallow as the source of fat rather than coconut oil. We also compared the effects of textured vegetable protein (TVP) and of a diet containing equal levels of beef protein and TVP [Kritchevsky et al, 1981]. The results are summarized in Table 5-X. Beef protein and casein were of equivalent atherogenicity and cholesterolemic effect but rabbits fed beef protein exhibited significantly lower HDL-cholesterol levels than did any of the other groups. Casein was the most triglyceridemic protein. The most striking finding was that a 1:1 mixture of beef and textured vegetable protein was no more atherogenic and only slightly more cholesterolemic than TVP alone. This observation holds important implications for the general diet since it suggests a means of obtaining the benefits of both animal and plant protein without the cholesterolemic effect of the former.

The level of dietary protein may also determine the level of atherogenicity. Newburgh and Clarkson [1923] found that rabbits fed 36% protein (as beef muscle) developed more severe atherosclerosis more rapidly than did rabbits fed 27% protein. Lofland et al [1966] compared the effects of high (30%) and low (8%) wheat gluten or casein-lactalbumin 85:15 in white Carneau pigeons fed four different dietary fats plus 30 mg of cholesterol/100 g of fat.

While effects on serum cholesterol levels were variable, birds fed the high protein consistently exhibited a greater prevalence of lesions (Table 5-XI). Strong and McGill [1967] fed baboons diets which were high or low in protein, high or low in cholesterol, and contained saturated or unsaturated fat. In most cases, the high-protein diets were more sudanophilic. Freyberg [1937] fed three groups of six rabbits diets containing 13.1% or 33.0% soy protein or 37.8% protein as gluten flour. He observed no atherosclerosis. Terminal serum cholesterol levels (of rabbits killed at varying times) were 13.1% soy, 79 ± 9 mg/dl; 33.0% soy, 82 ± 11 mg/dl; and 37.8% gluten, 111 ± 18 mg/dl. Freyberg challenged the findings of Newburgh and Clarkson [1923] concerning the atherogenicity of high levels of protein, not realizing that animal and vegetable proteins had different effects. We have compared the effects of 25% and 37.5% soy protein in rabbits and found no significant differences (Table 5-XII). Comparison of various levels of dietary casein is not available at this writing. Terpstra et al [1981] have fed rabbits diets containing 10%, 20%, or 40% casein. After 28 days, serum cholesterol levels (mg/dl) in the three groups were 120, 380, and 920, respectively. When the group fed 10% casein was then placed on the 40% casein diet for 21 days, cholesterol levels rose to 660 mg/dl. Reducing the casein content of the diet from 40% to 10% for 21 days caused cholesterol levels to drop to 280 mg/dl. Cholesterol levels of the rabbits maintained on 20% casein did not change.

Huff and Carroll [1980b] compared the effects of casein and soy protein on cholesterol oxidation and turnover in rabbits given 26-^{14}C-cholesterol. The

TABLE 5-X. Influence of Protein Source on Atherosclerosis in Rabbits*

	Protein			
	Beef (B)	TVP	B-TVP 1:1	Casein
Number	12	9	11	11
Serum lipids (mg/dl)				
Cholesterol	185 ± 24[a,b]	37 ± 4[a,c]	61 ± 6[b,d]	200 ± 18[c,d]
% HDL cholesterol	20.1 ± 2.1[a,b,c]	38.8 ± 5.1[a]	43.4 ± 4.2[b,d]	29.6 ± 3.6[c,d]
Triglycerides	60 ± 8[a]	59 ± 7[b]	70 ± 13	92 ± 10[a,b]
Phospholipids	92 ± 8[a,b,c]	67 ± 5[a,d]	70 ± 5[b,e]	126 ± 8[c,d,e]
Serum protein (g/dl)	5.53 ± 0.16	5.09 ± 0.21	5.24 ± 0.16	5.39 ± 0.20
Liver lipids (g/100 g)				
Cholesterol	0.77 ± 0.09[a,b]	0.28 ± 0.01[a,c]	0.38 ± 0.06[b,d]	0.97 ± 0.05[c,d]
Triglycerides	0.77 ± 0.08[a]	1.16 ± 0.39	0.72 ± 0.12	0.51 ± 0.08[a]
Phospholipids	1.46 ± 0.06[a]	1.18 ± 0.07[a,b,c]	1.58 ± 0.04[b,d]	1.41 ± 0.06[c,d]
Liver protein (g/100 g)	20.3 ± 0.7[a]	19.2 ± 1.2	19.9 ± 1.2	17.3 ± 1.00[a]
Atherosclerosis				
Arch	1.29 ± 0.23[a]	0.78 ± 0.12[b]	0.73 ± 0.10[a,c]	1.32 ± 0.22[b,c]
Thoracic	0.75 ± 0.12[a,b]	0.22 ± 0.09[a,c]	0.36 ± 0.12[b,d]	0.86 ± 0.12[c,d]

*Diets contained 40% sucrose, 25% protein, 14% beef tallow, 15% fiber (cellulose and soy fiber). Fed for 8 months. All data ± SEM.
[a-e]Values in horizontal line bearing the same letter are significantly different ($P < 0.05$).
After Kritchevsky et al [1981].

TABLE 5-XI. Influence of Protein Level on Cholesterolemia and Atherosclerosis in Pigeons

Fat	Casein-lactalbumin (85:15)		Heat gluten	
	Low[a]	High[b]	Low	High
Butter				
Cholesterol (mg/dl)	399	472	398	419
Incidence (%)	71	89	60	100
Corn oil				
Cholesterol	404	345	676	492
Incidence	56	100	56	78
Shortening				
Cholesterol	643	577	377	495
Incidence	55	89	58	85
Margarine				
Cholesterol	242	375	387	442
Incidence	42	73	36	60

[a]Low = 8% of calories.
[b]High = 30% of calories. Fats fed as 30% of calories.
After Lofland et al [1966].

TABLE 5-XII. Influence of 25% and 37.5% Soy Protein on Atherosclerosis in Rabbits

	Percent soy protein[a]	
	25.0	37.5
Number	11	10
Weight gain (g)	512 ± 93	1002 ± 131
Liver weight (g)	77 ± 6	85 ± 10
Relative liver weight	2.25 ± 0.12	2.12 ± 0.11
Serum lipids (mg/dl)		
Cholesterol	109 ± 25	57 ± 5
% HDL cholesterol	33.8 ± 4.5	44.1 ± 4.9
Triglycerides	45 ± 14	44 ± 6
Phospholipids	172 ± 30	180 ± 72
Serum protein (g/dl)	3.96 ± 0.14	3.65 ± 0.10
Liver lipids (g/100 g)		
Cholesterol	0.65 ± 0.09	0.55 ± 0.10
Triglycerides	2.71 ± 0.26	2.38 ± 0.17
Phospholipids	1.96 ± 0.08	1.97 ± 0.10
Liver protein (g/100 g)	18.4 ± 0.32	18.0 ± 0.45
Aortic atherosclerosis		
Arch	0.55 ± 0.18	0.50 ± 0.17
Thoracic	0.23 ± 0.141	0.15 ± 0.07

[a]Diets contained 40 or 27.5% sucrose, 15% cellulose, 14% coconut oil. Fed 8 months. All data ± SEM.
From Kritchevsky, unpublished observation.

study was conducted in rabbits fed diets high (15% butter) or low (1% corn oil) in fat. A group fed commercial ration was also included. The characteristics of cholesterol turnover are summarized in Table 5-XIII. Regardless of the dietary fat, cholesterol turnover in rabbits fed soy protein resembled that of rabbits fed commercial ration. Rabbits fed casein exhibited significantly larger body pools of cholesterol, higher cholesterol levels, and oxidized less cholesterol than rabbits fed soy protein. The rapidly metabolizing cholesterol body pool (A) was similar in rabbits fed soy protein or commercial ration, but the latter group exhibited a smaller pool B (slowly turning over) and oxidized significantly more cholesterol. Hermus [1975] compared cholesterol turnover in rabbits fed casein or a commercial ration and found the casein-fed rabbits to have a significantly larger pool A (but not pool B) and much slower turnover.

We have compared cholesterol absorption and excretion in rabbits fed the semipurified, casein-rich diet or a commercial ration (Table 5-XIV) [Kritchevsky et al, 1975]. Three days before termination of the experiment each rabbit was given an intraperitoneal injection of tritium-labeled cholesterol and carbon-labeled mevalonic acid. Thus ^3H radioactivity represented exogenous cholesterol and ^{14}C radioactivity represented endogenous cholesterol. Rabbits fed the commercial ration absorbed less cholesterol, exogenous or endogenous, as seen by the significantly lower levels of either ^3H or ^{14}C in liver and serum. Rabbits fed commercial ration excreted four times more ^3H and about 25% more ^{14}C in their feces. The rabbits fed the semipurified diet excreted more acidic steroid. The ratios of neutral/acidic fecal steroid in the two groups were as follows: For ^3H the ratios were semipurified 3.95, commercial 117.5; for ^{14}C the ratios were semipurified 1.36, commercial 30.9. Less cholesterol

TABLE 5-XIII. Influence of Dietary Protein on Cholesterol Oxidation and Turnover*

		Parameter		
Diet	M_A (mg)	M_B (mg)	Plasma cholesterol (mg/dl)	Cholesterol (mg/day) Oxidized to $^{14}CO_2$
Casein				
Low fat	1116 ± 99[a]	971 ± 112	335 ± 36[d]	13 ± 2[f]
High fat	1283 ± 195[b]	1392 ± 177[c]	289 ± 38[e]	20 ± 4[g]
Soy protein				
Low fat	726 ± 108[a]	677 ± 79	42 ± 5[d]	32 ± 2[f,h]
High fat	671 ± 79[b]	585 ± 79[c]	67 ± 7[e]	36 ± 2[g,i]
Commercial	784 ± 95	478 ± 107	63 ± 6	54 ± 5[h,i]

*M_A = mass, pool A; M_B = minimum mass, pool B; plasma cholesterol, mean over 42-day study. Rabbits fed 27–30% protein; 1% corn oil (low); 15% butter (high).
[a-i]Values bearing the same letter are significantly different ($P < 0.05$).
After Huff and Carroll [1980b].

TABLE 5-XIV. Influence of Diet on Cholesterol Metabolism in Rabbits (Data ± SEM)

	Diet[a]	
	Semipurified	Commercial
Number	9	5
Cholesterol		
Serum (mg/dl)	215 ± 17	93 ± 5*
Liver (mg/100 g)	225 ± 10	144 ± 4*
Recovery of radioactivity		
Serum ^{3}H (dpm × 10⁵)	4.03 ± 0.36	0.38 ± 0.11*
^{14}C (dpm)	27.57	ND[b]
Liver ^{3}H (dpm × 10⁶)	4.41 ± 0.52	1.34 ± 0.44*
^{14}C (dpm × 10⁴)	3.99 ± 0.59	1.05 ± 0.28
Feces		
Weight (g/3 days)	23 ± 9	98 ± 18*
Neutral steroid		
^{3}H (dpm × 10⁶)	3.75 ± 1.3	18.8 ± 10.4
^{14}C (dpm × 10³)	5.10 ± 1.8	10.8 ± 4.2
Acidic steroid		
^{3}H (dpm × 10⁶)	0.95 ± 0.22	0.16 ± 0.09*
^{14}C (dpm × 10³)q	3.76 ± 1.24	0.35 ± 0.11*

[a]Semipurified diet: 25% casein, 20% sucrose, 20% starch, 15% cellulose, 14% coconut oil. Fed 6 months. Rabbits given 1,2-^{3}H-cholesterol and 2-^{14}C-mevalonic acid 72 hr before termination of study; ^{3}H represents exogenous cholesterol, ^{14}C endogenous cholesterol.
[b]Not detectable.
*Significant, $P < 0.05$.
After Kritchevsky et al [1975].

was converted to acidic fecal products in rabbits fed commercial ration. Fumagalli et al [1978] made a similar observation.

We have established kinetic parameters of cholesterol metabolism in rabbits fed soy protein, soy protein plus lysine, or commercial ration [Czarnecki, 1982]. The rabbits (three per group) were injected intravenously with 4-^{14}C-cholesterol, and serum cholesterol specific activity was determined over a 57-day period. The computer analysis was carried out by Dr K.J. Ho [Ho and Taylor, 1973]. The results are shown in Table 5-XV. It is clear that the characteristics of cholesterol metabolism in rabbits fed soy protein plus lysine are similar to those observed by Huff and Carroll [1980b] in rabbits fed casein. Comparison of cholesterol kinetics in rabbits fed casein or casein plus arginine is in progress.

In summary, diets containing animal protein are generally more cholesterolemic and atherogenic for rabbits than diets containing vegetable protein. The protein effect can be modified by other dietary components such as fiber or type of fat. Diets containing animal and vegetable protein in equal proportions are no more cholesterolemic or atherogenic than those containing vegetable

TABLE 5-XV. Kinetic Parameters of Cholesterol Metabolism in Rabbits Fed Soy Protein, Soy Plus Lysine, or Commercial Ration

	Diet[a]		
	Soy protein	Soy + lysine	Commercial
Production rate (mg/day)	66	58	78
Mass, pool A (mg)	427	711	200
Mass, pool B (mg)	791	1260	690
Mean transit time (days)	18.4	33.7	11.8

[a]Semipurified diet containing 40% sucrose, 25% protein, 15% cellulose and 13% coconut oil and 1% corn oil. Protein was soy isolate or soy plus lysine to give lysine/arginine ratio of 2.0. Rabbits injected with 4-^{14}C-cholesterol (58–63 × 10^6 dpm) and disappearance of plasma cholesterol radioactivity followed for 57 days.

protein alone. Cholesterol turnover is slower, excretion diminished and body pools increased in rabbits fed animal protein. We have hypothesized that the lysine/arginine ratio of the protein determines its atherogenic effect. Casein and soy protein, the substances used as representative animal and vegetable proteins, have lysine/arginine ratios of about 2.0 and 0.9. Altering the lysine/arginine ratio of casein (by addition of arginine) does not affect cholesterolemia but reduces severity of atherosclerosis. Addition of lysine to soy protein to increase its lysine/arginine ratio causes increases in cholesterol level and greater severity of atherosclerosis. Feeding of proteins with similar lysine content but different lysine/arginine ratios results in levels of atherosclerosis which are correlated significantly with those ratios. Katan et al [1982] have reported that the addition of amino acids other than arginine (glycine, alanine) will also reduce atherogenicity of casein. The precise mode of action of the proteins and how it is affected by addition of single amino acids remains to be elucidated.

ACKNOWLEDGMENTS

This work was supported in part by grants HL-03299 and CA-09171 and a Research Career Award (HL-00734) from the National Institutes of Health; by a grant 59-2426-0-1-479 from the US Department of Agriculture, SEA; by the Mobil Foundation, Inc; and by the Commonwealth of Pennsylvania.

REFERENCES

Carroll KK, Hamilton RMG (1975): Effects of dietary protein and carbohydrate on plasma cholesterol levels in relation to atherosclerosis. J Food Sci 40:18–23.
Cittadini D, Pietropaolo C, DeCristofaro D, D'Ayjello M (1964): In vivo effect of L-lysine on rat liver arginase. Nature 203:643–644.
Cookson FB, Altschul R, Fedoroff S (1967): The effects of alfalfa on serum cholesterol and in

modifying or preventing cholesterol-induced atherosclerosis in rabbits. J Atheroscler Res 7: 69–81.

Cookson FB, Fedoroff S (1968): Quantitative relationships between administered cholesterol and alfalfa required to prevent hypercholesterolaemia in rabbits. Br J Exp Pathol 49:348–355.

Czarnecki SK (1982): Effects of dietary proteins on lipoprotein metabolism and atherosclerosis in rabbits. Ph.D. Dissertation, University of Pennsylvania, Philadelphia.

Czarnecki SK, Kritchevsky D (1979): The effect of dietary proteins on lipoprotein metabolism and atherosclerosis in rabbits. J Am Oil Chem Soc 56:388A.

Czarnecki SK, Kritchevsky D (1980): Effects of dietary protein on lipoprotein metabolism and atherosclerosis in rabbits. Fed Proc 39:342.

Enselme J, Cottet J, Fray G (1963): Etude de diverses influences alimentaires sur l'atherosclerose provoquee par une alimentation priovee de cholesterol. Extrait des Archives des Maladies du Coeur 56 (Suppl 3), Revue de l'atherosclerose 5:52–59.

Freyberg RH (1937): Relation of experimental atherosclerosis to diets rich in vegetable protein. Arch Intern Med 59:660–666.

Fumagalli R, Paoletti R, Howard AN (1978): Hypocholesterolaemic effect of soy. Life Sci 22: 947–952.

Hermus RJJ (1975): "Experimental Atherosclerosis in Rabbits on Diets With Milk Fat and Different Proteins." Wageningen, Netherlands: Centre for Agricultural Publications and Documentation.

Ho KJ, Taylor CB (1973): Female sex hormones: Effects on the kinetics of cholesterol metabolism in rabbits. Proc Soc Exp Biol Med 143:810–815.

Howard AN, Gresham GA, Jones D, Jennings IW (1965): The prevention of rabbit atherosclerosis by soya bean meal. J Atheroscler Res 5:330–337.

Huff MW, Carroll KK (1980a): Effects of dietary proteins and amino acid mixtures on plasma cholesterol levels in rabbits. J Nutr 110:1676–1685.

Huff MW, Carroll KK (1980b): Effects of dietary protein on turnover, oxidation and absorption of cholesterol and on steroid excretion in rabbits. J Lipid Res 21:546–558.

Huff MW, Hamilton RMG, Carroll KK (1977): Plasma cholesterol levels in rabbits fed low fat, cholesterol-free semipurified diets: Effects of dietary proteins, protein hydrolysates and amino acid mixtures. Atherosclerosis 28:187–195.

Hunter A, Downs CE (1945): The inhibition of arginase by amino acids. J Biol Chem 157: 427–446.

Jones JD (1964): Lysine-arginine antagonism in the chick. J Nutr 84:313–321.

Jones JD, Wolters R, Burnett PC (1966): Lysine-arginine electrolyte relationships in the rat. J Nutr 89:171–188.

Katan MB, Vroomen LHM, Hermus RJJ (1982): Reduction of casein-induced hypercholesterolemia and atherosclerosis in rabbits and rats by dietary glycine, arginine and alanine. Atherosclerosis 43:381–391.

Kritchevsky D (1967): Current concepts in the genesis of the atherosclerotic plague. In Best AN, Moyer JH (eds): "Atherosclerotic Vascular Disease." New York: Appleton-Century-Crofts, pp 1–7.

Kritchevsky D (1979): Vegetable protein and atherosclerosis. J Am Oil Chem Soc 56:135–146.

Kritchevsky D (1980): Dietary protein in atherosclerosis. In Noseda G, Lewis B, Paoletti R (eds): "Diet and Drugs in Atherosclerosis." New York: Raven Press, pp 9–14.

Kritchevsky D, Tepper SA (1965): Factors affecting atherosclerosis in rabbits fed cholesterol-free diets. Life Sci 4:1467–1471.

Kritchevsky D, Tepper SA (1968): Experimental atherosclerosis in rabbits fed cholesterol-free diets: Influence of chow components. J Atheroscler Res 8:357–369.

Kritchevsky D, Tepper SA, Czarnecki SK, Klurfeld DM (1982a): Atherogenicity of animal and vegetable protein. Influence of the lysine/arginine ratio. Atherosclerosis 41:429–431.

Kritchevsky D, Tepper SA, Czarnecki SK, Klurfeld DM, Story JA (1981): Experimental atherosclerosis in rabbits fed cholesterol-free diets. 9. Beef protein and textured vegetable protein. Atherosclerosis 39:169–175.

Kritchevsky D, Tepper SA, Czarnecki SK, Story JA, Marsh JB (1982b): Experimental atherosclerosis in rabbits fed cholesterol-free diets. 11. Corn, protein, wheat gluten and lactalbumin. Nutr Rep Int (In press).

Kritchevsky D, Tepper SA, Kim HK, Moses DE, Story JA (1975): Experimental atherosclerosis in rabbits fed cholesterol-free diets. 4. Investigation into the source of cholesterolemia. Exp Mol Pathol 22:11–19.

Kritchevsky D, Tepper SA, Story JA (1976): Experimental atherosclerosis in rabbits fed cholesterol-free diets. 6. Skim milk powder. Nutr Rep Int 13:207–211.

Kritchevsky D, Tepper SA, Story JA (1978): Influence of soy protein and casein on atherosclerosis in rabbits. Fed Proc 37:747.

Kritchevsky D, Tepper SA, Williams DE, Story JA (1977): Experimental atherosclerosis in rabbits fed cholesterol-free diets. 7. Interaction of animal or vegetable protein with fiber. Atherosclerosis 26:397–403.

Lofland HB, Clarkson TB, Rhyne L, Goodman HO (1966): Interrelated effects of dietary fats and proteins on atherosclerosis in the pigeon. J Atheroscler Res 6:395–403.

Meeker DR, Kesten HD (1940): Experimental atherosclerosis and high protein diets. Proc Soc Exp Biol Med 45:543–545.

Meeker DR, Kesten HD (1941): Effect of high protein diets on experimental atherosclerosis of rabbits. Arch Pathol 31:147–162.

Milner JA, Hassan AS (1981): Species specificity of arginine deficiency induced hepatic steatosis. J Nutr 111:1067–1073.

Milner JA, Perkins EG (1978): Liver lipid alterations in rats fed arginine-deficient diets. Lipids 13:563–565.

Milner JA, Prior RL, Visek WJ (1975): Effect of arginine deficiency on growth and intermediary metabolism in rats. J Nutr 104:1681–1689.

Moore JH (1967): The effects of the type of roughage in the diet on plasma cholesterol levels and aortic atherosis in rabbits. Br J Nutr 21:207–215.

Newburgh LH, Clarkson S (1923): The production of arteriosclerosis in rabbits by feeding diets rich in meat. Arch Intern Med 31:653–676.

Newman HAI, Zilversmit DB (1964): Accumulation of lipid and nonlipid constituents in rabbit atheroma. J Atheroscler Res 4:261–271.

Rudel LL, Lee JA, Morris MD, Felts JM (1974): Characterization of plasma lipoproteins separated and purified by agarose column chromatography. Biochem J 139:89–95.

Shore B, Shore V, Salel A, Mason D, Zelis R (1974): An apolipoprotein preferentially enriched in cholesterol ester-rich, very low density lipoproteins. Biochem Biophys Res Commun 58:1–7.

Strong JP, McGill HC (1967): Diet and experimental atherosclerosis in baboons. Am J Pathol 50:669–690.

Terpstra AHM, Harkes L, van der Veen FH (1981): The effect of different proportions of casein in semipurified diets on the concentration of serum cholesterol and the lipoprotein composition in rabbits. Lipids 16:114–119.

Weigensberg BI, Stary HC, McMillan GC (1964): Effect of lysine deficiency on cholesterol atherosclerosis in rabbits. Exp Mol Pathol 3:444–454.

Animal and Vegetable Proteins in Lipid
Metabolism and Atherosclerosis, pages 101–110
© *1983 Alan R. Liss, Inc., 150 Fifth Ave., New York, NY 10011*

6

Effects of Soy Protein on Cholesterol Metabolism in Swine

D.N. Kim, K.T. Lee, J.M. Reiner, and W.A. Thomas
Department of Pathology, Neil Hellman Building, Albany Medical College, Albany, New York 12208

I. INTRODUCTION

The objectives of this article are twofold: (1) to illustrate that soy protein and soy protein products are less hypercholesterolemic than casein [Kim et al, 1978; Kim et al, 1980a; Forsythe et al, 1980; Julius and Wiggers, 1979] and (2) to delineate underlying mechanisms leading to "hypocholesterolemia" in a swine model.

Our studies in swine were initiated as a follow-up of the reports on rabbits of Carroll and Hamilton [1975; Hamilton and Carroll, 1976] and on humans of Sirtori et al [1977]. These studies in rabbits used semipurified, low-fat diets without cholesterol while the diets in human studies contained varying amounts of fats and cholesterol.

We have demonstrated in young male Yorkshire swine [Kim et al, 1978; Kim et al, 1980a,b] (initial weight approximately 10 kg) fed a semipurified diet containing 40% fat (by calories) and 1 g cholesterol that a significant but smaller elevation in serum cholesterol concentrations results in 4–6 weeks when the protein in the diet is derived exclusively from soy protein as compared to casein (20% by calories for both proteins).

In our initial studies [Kim et al, 1978; Kim et al, 1980a] we have used a textuired soy protein product, Pro-Lean, (62.5% protein, Miles Laboratories) which was analogous to that used by Sirtori et al in a human study [Sirtori et al, 1977].

It had been alledged that hypocholesterolemia might have been due to the substance(s) contained in the textured soy protein product such as saponin [Potter et al, 1979], fibers [Helms, 1977], and other ingredients in small quantities. However, comparisons among purified soy protein products, Uncolored Pro-Lean (no extra salt added), and Pro-Lean, a textured soy protein product, have shown that all of these products are equally effective in keeping serum cholesterol levels low [Kim et al, 1978; Kim et al, 1980a]. Others also have suggested that saponin, fiber, and other substances may not have played an important role under the conditions of the experiments [Gatti and Sirtori, 1977; Kritchevsky, 1979a].

II. AMOUNT OF PROTEIN

The standard high-fat, high-cholesterol diet employed in the swine study was 20% protein (in calories) [Kim et al, 1978; Kim et al, 1980a]. In our previous studies [Kim et al, 1978; Kim et al, 1980b] a mixture of soy protein and casein of full doses or 1/2 soy protein and 1/2 casein (protein equivalent of the standard high-fat, high-cholesterol diet), were equally effective in keeping serum cholesterol lower than that with a casein diet.

In our latest study (Table 6-I), we have compared the effects on serum cholesterol values of casein or soy protein at 20% and 10% (by calories) levels. After six weeks of feeding a high-fat (40% by calories) high-cholesterol (1 g) diet, 10% soy protein diet (127 ± 27 mg/dl) was as effective as a 20% soy protein diet (113 ± 11 mg/dl) at keeping values low. However serum cholesterol levels in the group fed 10% casein (208 ± 18 mg/dl) were significantly less than in the 20% casein group (361 ± 52 mg/dl) but still significantly higher than in soy protein groups.

From these results, we can conclude that the hypercholesterolemic effect of casein is greater with a higher dose of casein. Soy protein is equally effective at both protein doses, 20% and 10% (by calories), at keeping serum cholesterol values low.

Therefore, the effectiveness of soy protein when a mixture of full doses of casein and soy protein were given was perhaps due to cancellation of hypercholesterolemic effect of casein by soy protein. A similar conclusion may be drawn for the case of one half/one half dose mixture of casein and soy protein.

III. AMINO ACID SUPPLEMENTATION

One of the leading hypotheses in regard to the mechanism of soy protein effects on cholesterol metabolism is based on the difference in the amino acid compositions of casein and soy protein [Huff et al, 1977; Yadar and Liener, 1977; Kritchevsky, 1979b]. Various single amino acids, combinations of amino acids, amino acid mixtures simulating casein or soy protein, protein

TABLE 6-I. Serum Cholesterol Concentrations and Body Weights in Swine Fed a Semipurified High-Fat (40% Calories), High-Cholesterol (1/g) Diet With Various Proteins for Six Weeks*

Group	Number of swine	Body weights[a] (mg/dl)		Serum cholesterol concentrations[b] (mg/dl)		
		Initial	Final	0 day	3 weeks	6 weeks
A. Soy protein	5	13.8 ± 0.7[c]	24.4 ± 0.9[d]	71 ± 4	89 ± 7	113 ± 11[d]
B. Casein	5	14.1 ± 0.7	21.3 ± 1.5[d]	70 ± 3	210 ± 25	361 ± 52[d]
C. Soy protein 1/2	5	14.6 ± 0.6	22.1 ± 0.7[d]	69 ± 2	103 ± 14	127 ± 26
D. Casein 1/2	5	14.2 ± 0.6	21.8 ± 0.7	78 ± 2	159 ± 17	208 ± 18
E. Egg albumin	5	14.1 ± 0.8	21.7 ± 1.3	72 ± 2	146 ± 22	153 ± 25
F. Gelatin, casein	5	14.2 ± 0.6	19.3 ± 0.7	66 ± 2	240 ± 28	335 ± 42

*Serum cholesterol concentrations were monitored weekly but initial, 3 and 6 weeks are presented here for simplicity.
[a]Body weights of A vs F, $P < 0.01$; all other comparisons, not significant (NS).
[b]In all groups serum cholesterol concentrations 0 day vs 6 weeks all significant $P < 0.001$. At 6 weeks: A vs C or E, NS; A vs B, D, or F $P < 0.01$; B vs C, D, E, or F $P < 0.01$; C vs D or F, $P < 0.05$; C vs E, NS; D vs E, NS; D vs F, $P < 0.05$; E vs F, $P < 0.01$.
[c]Mean ± standard error of mean.
[d]Number of swine for these values is 4.

hydrolysates, and protein mixtures to simulate the amino acid composition of soy proteins have been studied for their effects on serum cholesterol levels and atherosclerosis.

Based on the reports in rabbits [Hamilton and Carroll, 1976; Huff et al, 1977; Hermus and Dallinga-Thie, 1979], we have examined effects of supplementation of two amino acids, methionine and glycine. Methionine content is lower in soy protein, and supplementation of methionine to a soy protein diet (semipurified, low-fat, cholesterol-free) in rabbits is known to result in a moderate elevation of serum cholesterol [Hamilton and Carroll, 1976]. However, methionine supplementation to a soy protein diet (with high-fat, high-cholesterol) in humans did not affect the serum cholesterol levels [Sirtori et al, 1977]. Glycine was also reported to be hypocholesterolemic amino acid when supplemented to a casein diet in rabbits [Hermus and Dallinga-Thie, 1979] but its effects in man are not known.

In swine we have tested these two amino acids. Methionine supplementation in both casein and soy protein diets (high-fat, high-cholesterol) did not affect serum cholesterol levels [Kim et al, 1978]. Glycine supplementation had no effect on swine fed a soy protein diet, and it seemed to exacerbate hypercholesterolemia in swine fed a casein diet.

IV. DIFFERENT KINDS OF PROTEINS

Based on the studies of Carroll and colleagues [Hamilton and Carroll, 1976; Huff et al, 1977] and Hermus and Dallinga-Thie [1979] in rabbits, we have recently examined effects of egg albumin, gelatin, casein, and soy protein on serum cholesterol levels in swine when these proteins were supplied as the exclusive sources of dietary proteins in a high-fat, high-cholesterol diet (Table 6-I). After six weeks serum cholesterol levels of swine in each group (five swine) were for casein, soy protein, 1/2 gelatin + 1/2 casein, and egg albumin group, respectively, 361 ± 52 mg/dl, 113 ± 11 mg/dl, 335 ± 42 mg/dl, and 153 ± 25 mg/dl. The egg albumin diet group had a significantly lower serum cholesterol level than the casein group ($P < 0.01$), while the difference between the egg albumin diet group and the soy protein diet group was not significant. On the other hand, 1/2 gelatin + 1/2 casein in the diet was as hypercholesterolemic as the casein diet. Initially the swine in the gelatin diet group were given a full dose (20% protein by calories) of gelatin in the diet, but the direct observation and the record of daily food consumption revealed that swine did not eat the gelatin diet too well from the second day of the dietary regimen. By the end of the first week, it was firmly established that all swine in the group ate less than 50% of daily ration offered and the diet was modified to contain 1/2 gelatin and 1/2 casein mixture instead of full dose of gelatin. From the second week and throughout the experiment, swine in 1/2 gelatin and 1/2 casein diet group ate well and grew as well as the swine in the other groups.

As already mentioned, a less hypercholesterolemic effect of egg albumin was reported by Carroll et al [Hamilton and Carroll, 1976; Huff et al, 1977] in rabbits fed a semipurified, low-fat, cholesterol-free diet. Extrapolating these results to a situation where whole eggs are consumed it might be that hypercholesterolemia would not occur if proteins from other sources, particularly casein, were not in the diet. (The effect of egg lecithin should be assessed as well.)

In a previous study [Kim et al, 1974] in swine, eight medium-size boiled eggs (yolk and white) were added to a semipurified, high-fat diet with casein to see if there was any difference in hypercholesterolemic response between cholesterol from the eggs and crystalline cholesterol (2.4 g). No difference in serum cholesterol between the two groups was observed. Eight eggs supplied approximately 50 g of protein and the semipurified diet also contained 130 g of casein. Further exploration of the combination of various proteins is indicated.

V. EFFECTS OF AMOUNTS OF DIETARY CHOLESTEROL

In a recent study, we compared [Kim et al, 1982] the effects of varying amounts of dietary cholesterol on serum cholesterol concentrations in two semipurified, high-fat diets (40% fat by calories) with either soy protein or casein (20% by calories) as the sources of protein. Five groups of swine (five in each group) were initially fed mash for two weeks, then in succession all were given a soy protein diet for four weeks, mash for one week, and a casein diet for four weeks. The mash diet did not contain cholesterol. During soy protein- and casein-diet periods each group was given daily cholesterol of 0.5, 1, 2, 4, and 8 g, respectively. During the soy protein-diet period serum cholesterol levels were low in the four groups fed daily dietary cholesterol of 0.5, 1, 2, and 4 g. The group that was given 8 g cholesterol daily exhibited hypercholesterol- emia similar to that observed in swine fed casein diet with 1 g of dietary choles- terol daily in the previous studies. After four weeks of soy protein diets all five groups were switched to a mash diet without cholesterol for one week. Serum cholesterol concentration of all swine had returned to their initial mash diet levels at the end of the week. This was followed by casein diet for four weeks: Five groups of swine received the same amount of cholesterol that was given during soy protein diet –0.5, 1, 2, 4, and 8 g daily. At the end of four weeks, serum cholesterol concentrations were increased in proportion to the dietary cholesterol intakes up to 4 g; however, no further increase in serum cholesterol was observed in groups receiving 8 g dietary cholesterol.

These results led us to conclude that a soy protein diet fed to swine can handle a wide range of dietary cholesterol intakes up to 4 g (or thereabouts), but 8 g was beyond the limits the system could handle.

VI. HYPOCHOLESTEROLEMIC MECHANISM

Our approach to elucidate the hypocholesterolemic mechanism of soy pro- tein is based on the traditional cholesterol balance concept (Fig. 6-I). Serum cholesterol concentrations represent the most readily accessible body-com- partment cholesterol pool. This compartment will equilibrate with other body tissue pool in time and this can be confirmed in experimental animals by direct measurements.

Assuming for a moment that changes in serum cholesterol levels are a reflec- tion of changes in the overall retention of body cholesterol, we can then say that a soy protein diet induces smaller retention of cholesterol than a casein diet.

In a steady-state condition, since there is no retention (change) in body cho- lesterol, the following relation can be written:

$$\text{Dietary intake (D) + synthesis (S) = excretion (E)} \tag{1}$$

In growing swine, or in swine where body cholesterol content is expected to change,

$$\text{Dietary intake (D) + synthesis (S) = retention (R) + excretion (E)} \tag{2}$$

The relation in Equation 2 represents a universal condition. If $R = 0$, the swine is in steady state condition and Equation 1 applies.

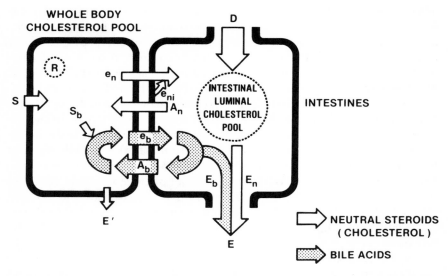

Fig. 6-I.   Various parameters of cholesterol balance: D, dietary cholesterol intake; S, whole-body cholesterol synthesis; R, cholesterol retention; E, total fecal steroid excretion; Sb, bile acid synthesis; E_n, E_b, fecal excretions, neutral steroid and bile acids; A_n, cholesterol absorption; A_b, bile acid absorption; e_b, endogenous bile acid excretion*; e_n, endogenous neutral steroid excretion from bile; e_{ni}, endogenous neutral steroid excretion from intestinal epithelia**; E', neutral steroid excretion from body through the routes other than feces, ie, skin, urine and others. (*Endogenous excretion is defined as neutral or acidic steroid that is delivered into the intestinal lumen. **Intestinal epithelial cell turnover and direct loss from the intestine.)

All of these parameters can be measured directly in experimental animals. Further breakdown of these parameters into absorption, endogenous excretion, and acidic and neutral steroids can generate valuable information.

We have measured (1) fecal steroid excretion, (2) tissue cholesterol concentration, (3) whole-body cholesterol synthesis, (4) hepatic 3-hydroxy-3-methylglutaryl coenzyme A (HMG-CoA) reductase activity, (5) dietary cholesterol absorption (in a few animals), and (6) serum cholesterol concentrations [Kim et al, 1978; Kim et al, 1980a]. We have demonstrated increases in fecal excretions of both neutral and acidic steroids in a soy protein diet period and increased during casein diet period. Tissue cholesterol concentration reflected serum cholesterol levels. No paradoxical accumulation of cholesterol in the tissue occurred. Whole-body cholesterol synthesis and hepatic microsomal HMG-CoA reductase activities in both groups were maximally inhibited as compared to activities in mash-fed swine, and cholesterol synthesis in both groups was the same.

Therefore, from Equation 2,

$$R = (D + S) - E \tag{3}$$

Under the given experimental condition, $(D + S)$ is constant and R is determined directly by E. As E increases, retention will decrease.

Now let us examine what constitutes excretion (E). E consists of both acidic (bile acids) and neutral steroids. The sole source of bile acids excreted in feces is the bile. Neutral steroids are derived from at least three sources: (1) dietary cholesterol unabsorbed, (2) endogenously excreted cholesterol in the bile unabsorbed, and (3) desquamated intestinal epithelium or directly from the intestinal mucosal cells. Since we may tentatively assume that the turnover of the intestinal epithelium is the same in both soy protein and casein group, we shall consider the first two sources: (1) dietary cholesterol and (2) endogenously excreted cholesterol, which comes from cholesterol in the bile. We will assume that endogenous cholesterol and dietary cholesterol will be sufficiently mixed and the intestine will not distinguish them.

Then, the amount of neutral steroid excreted in the feces will be determined by difference between total amount of the intestinal luminal cholesterol pool and amount of cholesterol adsorbed. Under the conditions of the experiment, dietary cholesterol intake (D) is constant and the same in both groups. Therefore, differences in neutral steroid excretion between the two groups must come from differences in absorption and/or in quantity of endogenous cholesterol excretion.

The amount of bile acid excreted in feces is dependent upon the difference between the amount of bile acid entering the intestine through bile excretion and the amount of bile acids absorbed. Normally only a small fraction of bile acids entering the intestine is unabsorbed and excreted in the feces. Small but consistent increases in bile acid excretion during soy protein diet period may have resulted from increased endogenous excretion of bile acids and/or reduced absorption of bile acids.

Thus reduced absorption and/or increased endogenous excretion of both neutral and acidic steroids appear to be the determining parameters that can explain the reduction of lesser increase in body cholesterol pool in the soy protein diet group as compared with the casein diet group.

We have not yet studied in depth which one of the parameters, absorption or endogenous excetion, is the determining factor for the explanation of hypocholesterolemic action of soy protein. In a few swine, we have measured dietary cholesterol absorption. The average absorption value for a soy protein group was lower than that for a casein group but the difference was not significant, perhaps because of group comparisons or the small number of swine in the study.

Table 6-II. Total Fat Excretion (g/day)* in a Crossover Study With a High-Fat, High-Cholesterol, Semipurified Diet With Either Soy Protein or Casein as the Source of Protein

Group[a]	Swine number	Period I Soy protein (g/day)	Period III Casein (g/day)	Change g	Change %
A	1	14.6	4.7	−9.9	−68
	2	10.5	4.9	−5.6	−53
	3	11.7	4.9	−6.8	−58
	4	24.2	5.2	−19.0	−79
	5	14.7	5.4	−9.3	−63
		15.1 ± 1.40[b]	5.0 ± 0.12	−10.1 ± 2.35	−64 ± 4.3
		↑ ($P < 0.01$) ↑			

Group	Swine number	Period I Casein (g/day)	Period III Soy protein (g/day)	Change g	Change %
B	6	7.5	7.9	0.4	6
	7	11.9	8.1	−3.8	−32
	8	6.6	8.9	2.3	34
	9	13.3	13.0	−0.3	−1
	10	12.9	15.2	2.3	18
		10.4 ± 1.41	10.6 ± 1.48[c]	0.2 ± 1.12	5 ± 10.99
		↑ (NS) ↑			

*Fat is extracted by the method of Folch; dried under nitrogen to a constant weight.

[a]Group A was given soy protein diet and group B was given casein diet initially for 4 weeks. During the last week of 4 week regimen, feces were collected and this week was designated as Period I. The diets were then crossed over and feces were collected for 2 weeks: the first week after crossover was designated Period II (data for this period not presented for simplicity), and the second as Period III.

[b]Mean ± standard error of mean.

[c]Difference between groups A and B during period III is significant ($P < 0.01$).

From Kim et al [1980a].

Recently we have analyzed the fat content of fecal samples from a previous study [Kim et al, 1980a] in which fecal steroid excretion was measured in crossover feeding of soy protein and casein diet (Table 6-II). Briefly, two groups of swine (five in each) weighing approximately 10 kg at the outset were given high-fat, high-cholesterol (1 g) diets either with soy protein or casein as the sole source of protein. After four weeks on given diets, the diets were exchanged and swine were fed for another two weeks. Fecal measurements were obtained from a one-week pool of daily fecal collections immediately prior to crossover (period I) and during the second week after crossover (period III; period II is the first week after changing the diet).

In the group that was given soy protein diets initially for four weeks and a casein diet later, fecal fat output was 15.1 ± 2.4 g/day and decreased to 5.0 ± 0.1 g/day during the second week (period III) after the diet was changed to casein diet ($P < 0.01$) the average reduction was 10.1 ± 2.4 g/day, or 64% ± 4%.

In the other group which was given a casein diet for four weeks before the casein diet was changed to soy protein diet, fecal fat excretion during the week before changing the diet (period I) was 10.4 ± 1.4 g/day. The average value during period II (the second week of soy protein diet period) was 10.6 ± 1.5 g/day. This represented virtually no change in fat excretion. The evaluation of individual swine indicated, however, that excretion increased in three swine and reduced in two swine. This effect was in sharp contrast to what we found in the other group in which a 50% reduction was observed when the diet was changed from soy protein to casein. For group comparisons, significant differences were observed during period III when a greater amount of fat was excreted in a group fed soy protein diet as compared to casein diet group.

Although the quantity of fat excreted represented a small fraction of total fat ingested, the consistency in which fat excretion has changed depending upon diet, strongly suggested that interference in fat absorption occurred with soy protein diet. This appears to further support the idea that effects on absorption of cholesterol underlie hypocholesterolemic mechanisms of soy protein diet in swine.

VII. SUMMARY

When young male Yorkshire swine are fed a semipurified, high-fat, high-cholesterol diet with casein as a source of protein, elevation of serum cholesterol concentration results. The same diet with soy protein instead of casein will raise serum cholesterol significantly less; and in crossover studies a definite hypocholesterolemic effect is observed.

The hypocholesterolemic effect of soy protein appears to be the result of reduction in cholesterol absorption and/or increase in endogenous cholesterol excretion. The underlying mechanism is not known.

VIII. REFERENCES

Carroll KK, Hamilton RMG (1975): Effects of dietary protein and carbohydrate on plasma cholesterol levels in relation to atherosclerosis. J Food Sci 40:18–23.

Forsythe WA, Miller ER, Hill GM, Romsos DR, Simpson RC (1980): Effects of dietary protein and fat source on plasma cholesterol parameters, LCAT activity and amino acid levels and on tissue lipid content of growing pigs. J Nutr 110:2467–2479.

Gatti E, Sirtori CR (1977): Soybean-protein diet and plasma cholesterol. (Reply) Lancet I 8015: 805–806.

Hamilton RMG, Carroll KK (1976): Plasma cholesterol levels in rabbits fed low-fat and low cholesterol diets. Effects of dietary proteins, carbohydrates, and fiber from different sources. Atherosclerosis 24:47–62.

Helms P (1977): Soybean protein diet and plasma cholesterol. Lancet I 8015:805.

Hermus RJJ, Dallinga-Thie GM (1979): Soya, saponins, and plasma-cholesterol. Lancet II 8132:48.

Huff MW, Hamilton RMG, Carroll KK (1977): Plasma cholesterol levels in rabbits fed low-fat, cholesterol-free, semi-purified diets: Effects of dietary proteins, protein hydrolysates and amino acid mixtures. Atherosclerosis 28:187–195.

Julius AD, Wiggers KD (1979): Effect of infant formulas and source of fat and protein on blood and tissue cholesterol in weanling pigs. Fed Proc 38:774.

Kim DN, Lee KT, Reiner JM, Thomas WA (1974): Restraint of cholesterol accumulation in tissue pools associated with drastic short-term lowering of serum cholesterol levels by clofibrate or cholestyramine in hypercholesterolemic swine. J Lipid Res 15:326–331.

Kim DN, Lee KT, Reiner JM, Thomas WA (1978): Effects of soy protein product on serum and tissue cholesterol concentrations in swine fed high-fat, high-cholesterol diets. Exp Mol Pathol 29:385–399.

Kim DN, Lee KT, Reiner JM, Thomas WA (1980a): Increased steroid excretion in swine fed high-fat, high-cholesterol diet with soy protein. Exp Mol Pathol 33:25–35.

Kim DN, Lee KT, Reiner JM, Thomas WA (1980b): Studies of hypocholesterolemic effect of soy protein in swine. In Fumagalli R, Kritchevsky D, Paoletti R (eds): "Drugs Affecting Lipid Metabolism." The Netherlands: Elsevier/North-Holland Biomedical Press, pp 347–353.

Kim DN, Lee KT, Reiner JM, Thomas WA (1982): Hypolipidemic action of soy protein in swine. In Noseda G, Fragiacomo C, Fumagalli R, Paoletti R (eds): "Lipoproteins and Coronary Atherosclerosis." The Netherlands: Elsevier Biomedical Press B.V., pp 265–270.

Kritchevsky D (1979a): Soya, saponins, and plasma-cholesterol. Lancet I 8116:610.

Kritchevsky D (1979b): Vegetable protein and atherosclerosis. J Am Oil Chem Soc 56:135–140.

Potter JD, Topping DL, Oakenfull D (1979): Soya, saponins and plasma-cholesterol. Lancet I 8109:223.

Sirtori CR, Agrade E, Conti F, Montero O (1977): Soybean-protein diet in the treatment of type II hyperlipoproteinemia. Lancet I:275–277.

Yadar NR, Liener IE (1977): Reduction of serum cholesterol in rats fed vegetable protein or an equivalent amino acid mixture. Nutr Rep Int 16:385–389.

*Animal and Vegetable Proteins in Lipid
Metabolism and Atherosclerosis, pages 111–134
© 1983 Alan R. Liss, Inc., 150 Fifth Ave., New York, NY 10011*

7

Influence of Human Diets Containing Casein and Soy Protein on Serum Cholesterol and Lipoproteins in Humans, Rabbits, and Rats

Joop M.A. van Raaij, Martijn B. Katan, and Clive E. West
Department of Human Nutrition, Agricultural University, De Dreijen 12, 6703 BC Wageningen, The Netherlands

I. INTRODUCTION

In animal studies, dietary proteins derived from animal sources are generally found to be hypercholesterolemic when compared with proteins from plant sources [Carroll, 1978; Kritchevsky, 1979; Terpstra et al, 1982b]. Epidemiological studies [Stamler, 1979] and nutritional studies in vegetarians [Hardinge and Stare, 1954; Sacks et al, 1975; Burslem et al, 1978] have also suggested a relation between intake of animal protein and serum cholesterol. Such epidemiological observations, however, should be interpreted with caution as differences in other nutritional factors may be present. In humans, only a few controlled trials relating dietary protein to serum cholesterol have been carried out, and the results are conflicting [Anderson et al, 1971; Carroll et al, 1978; Sirtori et al, 1979; Shorey and Davis, 1979; Descovich et al, 1980; Vessby et al, 1980; Holmes et al, 1980; Bodwell et al, 1980; Wolfe et al, 1981].

We have recently investigated, in strictly controlled dietary studies, the effects of diets containing casein and soy protein on the concentration of serum cholesterol and lipoproteins in large groups of young students and middle-aged subjects. During our studies with humans, duplicate portions of the test diets were collected and were later fed to rabbits and rats. In this paper the results of the human and animal studies are discussed. Some of the results have been published previously [van Raaij et al, 1981, 1982].

II. MATERIALS AND METHODS

A. Human Subjects, Animals, and Diets

In the experiments to be described, three principal sources of protein have been used: casein; soy protein isolate, which is the purest form of soy protein commercially available; and soy protein concentrate, which contains more dietary fiber (Table 7-I). Six studies have been carried out and they can be divided into two groups on the basis of the diets fed (Table 7-II; Fig. 7-I).

TABLE 7-I. Composition of the Protein Preparations Used*

| | | Soy isolate | | |
	Caseinate[a]	UNISOL NH 70[b]	PP500E, PP610[c]	Soy concentrate[d]
Protein[e]	92.4	84.6	79.7	57.2
Moisture	3.0	3.1	5.5	5.9
Fat	0.3	0.1	0.5	1.6
Ash	4.2	3.5	4.0	6.4
Carbohydrates	0.2[f]	8.7[g]	10.3[g]	28.9[g]

*All values are g/100 g.

[a]Calcium and sodium caseinate (spray dried, bland), DMV Milk Industries, 5460 BA Veghel. Data expressed as mean of the values for calcium and sodium caseinate.

[b]UNISOL NH 70, UNIMILLS BV, 3330 AA Zwijndrecht (used in I-Human and I-Rabbit).

[c]Soy protein isolate PP500E and PP610, Purina Protein Europe, B-1050 Brussels, Belgium (used in II-Human, II-Rabbit, II-Rat and II-Rabbit-SP). Data expressed as mean of the values for PP500E and PP610.

[d]Soy protein concentrates Unico (powder) and Dubit (textured, prepared from Unico), UNIMILLS BV. Data expressed as mean of the values for Unico and Dubit.

[e]Kjeldahl nitrogen-to-protein factors of 6.38 for casein and 5.70 for soy protein have been used.

[f]Lactose; data provided by manufacturer.

[g]By difference.

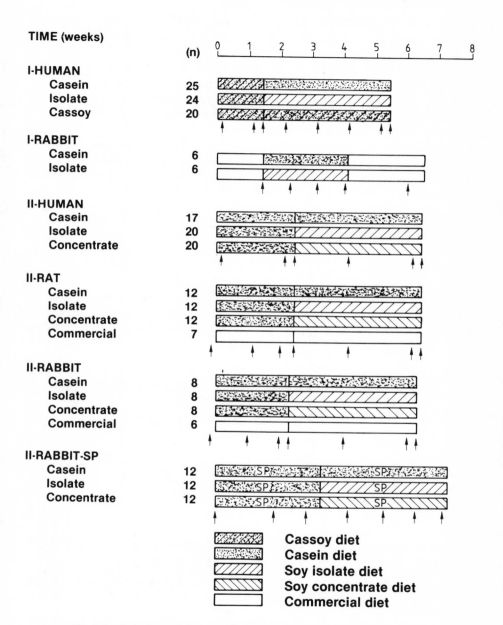

Fig. 7-I. Experimental designs. Blood sampling is indicated by arrows.

TABLE 7-II. Composition of the Diets Used (g per 100 g Dry Matter)

Experiment	Period and diet[a]	Protein[b]	Fat				Carbohydrates		Dietary fiber	Ash	Cholesterol	Fyto-sterols
			Total	Saturated	Mono-unsaturated	Poly-unsaturated	Total	Sugars				
I-Human	Control period											
I-Rabbit	Cassoy	18.4c	21.2	8.4	9.2	3.6	41.3	14.4	9.2	4.0	0.08	0.11
	Commercial	21.3	4.8	e	e	e	55.4	e	10.6f	7.9	0	e
	Test period											
	Cassoy	17.6c	21.3	8.1	9.4	3.8	42.1	15.9	8.6	3.7	0.08	0.12
	Casein	18.4c	22.1	8.1	9.9	4.1	42.9	17.4	9.1	3.7	0.07	0.12
	Soy isolate	17.4c	22.2	8.0	10.1	4.1	41.8	16.5	8.6	3.6	0.07	0.12
II-Human	Control period											
II-Rat	Casein	20.2c	20.6	8.4	7.7	4.5	46.2	16.0	7.8	4.5	0.07	0.07
II-Rabbit	Test period											
	Casein	20.1c	20.7	8.3	7.6	4.8	45.7	15.7	8.2	4.6	0.07	0.07
	Soy isolate	19.7c	19.6	7.6	7.2	4.8	45.2	15.4	10.5	4.3	0.07	0.07
	Soy concentrate	19.7c	18.4	7.2	6.6	4.6	42.9	14.0	12.8	4.6	0.06	0.07
	Commercial (Rat)	25.2	7.3	e	e	e	56.7	e	4.7f	6.1	0	e
	Commercial (Rabbit)	21.3	4.8	e	e	e	55.4	e	10.6f	7.9	0	e

II-Rabbit-SP Control period											
Casein	21.6[d]	5.3	4.0	0.5	0.7	53.4	27.2	10.5	[e]	0	[e]
Test period											
Casein	21.6[d]	5.3	4.0	0.5	0.7	53.4	27.2	10.5	[e]	0	[e]
Soy isolate	19.1[d]	5.3	4.0	0.5	0.8	53.1	27.0	12.8	[e]	0	[e]
Soy concentrate	16.8[d]	5.1	3.6	0.6	0.9	47.7	24.2	17.7	[e]	0	[e]

[a]The test diets used in all experiments, except in the control period of I-Rabbit and throughout II-Rabbit-SP, were conventional human diets. These diets were freeze-dried before feeding to the animals. The composition of these diets was determined by chemical analysis [van Raaij et al, 1981, 1982]. In II-Rabbit-SP the protein preparations were incorporated into semipurified pelleted diets. The composition of these diets was calculated on basis of the ingredients used (in g/1,000, 1,006, and 1,084 g for the casein, isolate, and concentrate diet, respectively): casein, 210, or soy protein isolate, 216, or soy protein concentrate, 295; maize starch, 280; dextrose, 210; saw dust, 120; coconut oil, 40; soybean oil, 10; molasses, 50; vitamin premix, 12; mineral premix, 10; CaHPO₄.2H₂O, 29; NaCl, 6; MgCO₃, 3; MgO, 2; and KHCO₃, 18. As soy protein is deficient in methionine, and as the soy preparations contained more NaCl than casein, 2 g of NaCl was replaced by 2.1 and 1.4 g DL-methionine in the isolate and concentrate diets, respectively. The composition of the vitamin and mineral premixes has been described earlier [Katan et al, 1982]. In addition to the human diets, pelleted commercial diets were used in I-Rabbit, II-Rabbit, and II-Rat. Data on the composition of these diets were provided by manufacturers.

[b]The proportion of protein as test protein was 65% in I-Human and I-Rabbit, 60% in II-Human, II-Rat, and II-Rabbit, and 100% in II-Rabbit-SP.

[c]Kjeldahl nitrogen to protein factor of 6.25 was used.

[d]Kjeldahl nitrogen to protein factors of 5.70 and 6.38 were used for casein and soy protein, respectively.

[e]No data available.

[f]Crude fiber.

The first group of studies (van Raaij et al [1981] referred to in this paper as group I) consisted of a study with humans (I-Human) and a study with rabbits (I-Rabbit). In I-Human, 69 young healthy students aged 18 to 28 years were fed for 38 days on diets containing 13% of energy as protein of which 65% was replaced by protein from casein, from soy protein isolate, or from a 2:1 mixture of casein and soy protein isolate. After a control period of ten days during which all the subjects received the casein-soy diet, 20 subjects continued on this diet for a test period of 28 days, 25 subjects switched to the casein diet, and the remaining 24 subjects switched to the soy diet. Throughout the experiment duplicate portions of the diets were collected, and these were later homogenized in order to carry out a study with rabbits (I-Rabbit). Twelve male New Zealand white rabbits were maintained on a commercial diet (Cunicon I, Trouw and Co, 3881 LP Putten) until 3 months of age (control period-1). Half were transferred to the homogenized casein diet and half were transferred to the soy isolate diet for a test period of 2½ weeks before both groups were transferred back to the commercial diet for 2½ weeks (control period-2). The homogenate was supplied freeze-dried and pelleted during the first 12 days of the test period and as a wet mash during the remaining six days.

The second group of studies (group II) included a study with humans (II-Human), one study with rats (II-Rat) and two studies with rabbits (II-Rabbit and II-Rabbit-SP). In II-Human [van Raaij et al, 1982], 57 healthy subjects aged 29 to 60 years were fed for 45 days on diets containing 16% of energy as protein of which 60% was replaced by protein from casein, from soy protein isolate, or from soy protein concentrate. After a control period of 17 days during which all the subjects received the casein diet, 17 subjects continued on this diet for a test period of 28 days, 20 subjects switched to the soy isolate diet, and the remaining 20 subjects switched to the concentrate diet. As during the first study with humans, duplicate portions of the diets were collected, homogenized and freeze-dried in order to carry out a study with male New Zealand white rabbits (II-Rabbit) and also with female lean Zucker strain rats (II-Rat). The experimental design of these two studies was essentially the same as II-Human. Before the control periods, the rats and rabbits were maintained on commercial rat and rabbit pellets, respectively (Trouw and Co). In addition a further experiment with rabbits (II-Rabbit-SP) was carried out according to a similar design, but this time we used semipurified diets in which 100% of the protein was supplied by the same preparations of casein, soy isolate, and soy concentrate as used in II-Human, II-Rat, and II-Rabbit. At the beginning of the control periods, the rabbits and rats in the second group of experiments were 10–13 and 4 weeks of age, respectively.

For the two human experiments, food records and chemical analysis of the diets indicated that within each study there were essentially no differences between the experimental diets except for the type of protein and/or the amount of nonprotein material derived from the protein preparations used. Apart from 120 kcal per day in I-Human and 240 kcal per day in II-Human, all of the food eaten was daily supplied to the subjects in amounts appropriate to each individual's energy requirements. The composition of the diets is given in Table 7-II, but more details about the diets have been described previously [van Raaij et al, 1981, 1982].

In the animal studies water was provided ad libitum. In I-Rabbit and II-Rat, the casein and soy diets were provided ad libitum, while in II-Rabbit and II-Rabbit-SP, the diets were fed on a restricted basis to supply the rabbits in each experiment with equal amounts of protein. The animals were maintained as described by Katan et al [1982] except that the rats as well as the rabbits were housed individually. The individual body weights were recorded weekly.

B. Sampling of Blood and Liver and Biochemical Analysis

Blood samples were collected throughout the experiments as shown in Figure 7-I. Blood was taken after an overnight fast as described earlier for humans [van Raaij et al, 1981], rabbits [Terpstra and Sanchez-Muniz, 1981] and rats [Terpstra et al, 1982a]. Serum was obtained by low-speed centrifugation, and all the serum samples were assayed for total cholesterol.

At the end of both periods, in both human studies and in II-Rat and II-Rabbit-SP, the lipoproteins of the serum were isolated by density-gradient ultracentrifugation using a modification [Terpstra et al, 1981b] of the method described by Redgrave et al [1975]. An SW50 rotor (Beckman Inc, Palo Alto, CA 94304, USA) was used for the separation of lipoprotein classes from 1-ml serum samples, as in II-Rat, and an SW41 rotor was used for the separation from 2-ml samples, as in all other studies. The lipoproteins in the gradient were visualized by prestaining the serum with Sudan black prior to ultracentrifugation. Lipoprotein fractions were removed either by tube-slicing (I-Human) or by aspiration (all other studies). In the human studies the following lipoprotein fractions were collected from individual sera: in I-Human — very low density lipoprotein (VLDL) (d < 1.015 g/ml), low-density lipoprotein (LDL) (1.015 < d < 1.060), and high-density lipoprotein (HDL) (d > 1.060 g/ml); and in II-Human — VLDL (d < 1.006 g/ml), LDL + sinking pre-beta-lipoproteins (1.006 < d < 1.075), and HDL (d > 1.075). In I-Rat, the serum samples were pooled per group at the end of the control period and again at the end of the test period. Nine lipoprotein fractions were collected, eight of which floated above the following densities (d, g/ml), respectively: 1.016, 1.032, 1.045, 1.060, 1.078, 1.102, 1.128, and 1.146, while the ninth fraction had a d > 1.146 g/ml. In II-Rabbit-SP analyses were carried out on pooled samples with similar cholesterol concentrations. At the end of the control period, six pools each of six sera were formed and at the end of the test period there were four pools within each group with each pool consisting of three sera. Three lipoprotein fractions were collected: VLDL (d < 1.006), LDL (1.006 < d < 1.063), and HDL (d > 1.063 g/ml).

At the end of experiment II-Rabbit-SP, ie, after five weeks on the test diets, the concentration of cholesterol and the proportion present as the ester was measured in the livers of the rabbits. The animals were first stunned by a blow to the head and then killed by severing the major blood vessels in the neck and draining the blood from the body. The livers were removed immediately, dried off, and weighed. The liver was then homogenized with a Potter-Elvehjem homogenizer and lipids were extracted from a 10-g aliquot [Folch et al, 1957]. We measured total cholesterol in one part of the extract with Liebermann-Burchard reagent [Huang et al, 1961] after alkaline hydrolysis. Another part of the extract was used for separating and quantitating free and esterified cholesterol by thin-layer chromatography [West and Rowbotham, 1967].

In both human studies serum cholesterol was measured with the reagent of Huang et al [1961] with strict standardization [Katan et al, 1982]; animal sera and lipoprotein fractions were assayed for cholesterol by an enzymatic method [Röschlau et al, 1974] using a kit (Catalase kit, no. 124087, Boehringer-Mannheim GmbH, West Germany). Apolipoprotein-B (I-Human and II-Human) and apolipoprotein A_I (II-Human) were measured in whole serum by rocket immunoelectrophoresis [Laurell, 1972] largely as described previously [Brussaard et al, 1980]. Rabbit antibody against apolipoprotein A_I was kindly donated by Dr. P. Demacker.

The response to the test diet was calculated per subject or animal as the change from the end of the control to the end of the test period. Differences in diet effects were examined by comparing the mean responses of the groups by unpaired two-tailed t-tests [Snedecor and Cochran, 1967].

III. RESULTS

A. Human Experiments

The concentrations throughout experiments I-Human and II-Human of cholesterol in whole serum, HDL, and LDL and of apolipoproteins A_I (II-Human only) and B are given in Tables 7-III and 7-IV, and in Figure 7-II. In both studies, there was no clear difference in response with respect to serum total cholesterol between the casein and soy isolate groups. In both studies a decline in LDL cholesterol and an increase in HDL cholesterol was observed on the soy isolate diet when compared with the casein diet, but only in I-Human the difference with respect to LDL cholesterol reached statistical significance (Fig. 7-II; van Raaij et al [1981, 1982]).

In I-Human an increase in apolipoprotein-B and a decline in the LDL cholesterol/apolipoprotein-B ratio was observed on the isolate diet when compared with the casein diet, suggesting that the decline in LDL cholesterol on the isolate diet had not been caused by a decline in the number of LDL particles, but in the composition of the LDL. These findings are confirmed by the changes in the density of LDL [van Raaij et al, 1981].

In II-Human a decline in apolipoprotein-B, an increase in apolipoprotein-A_I, and no clear differences in LDL cholesterol/apo-B and HDL cholesterol/apo-A_I ratios were observed on the soy isolate diet when compared with casein. These results suggest that the decline in LDL cholesterol and the increase in HDL cholesterol were caused by a decline in the number of LDL particles and an increase in the number of HDL particles, rather than by changes in the lipoprotein composition. The correlation coefficients between LDL cholesterol and serum apo-B at the end of the control and test period were $r =$

TABLE 7-III. The Effect of Casein and Soy Protein Diets on Serum Cholesterol Concentrations in Humans, Rabbits, and Rats

Experiment	Group[a]	n	Serum cholesterol concentration (mg/dl, mean ± SD[b])			
			Initial	End control period	End test period	Change over test period
I-Human	Casein	25	162 ± 32	152 ± 27	149 ± 24	− 3 ± 14
	Isolate	24	159 ± 26	153 ± 23	150 ± 23	− 3 ± 10
	Cassoy	20	160 ± 27	153 ± 24	150 ± 25	− 3 ± 13
I-Rabbit	Casein	6	c	30 ± 7	120 ± 65	+89 ± 74
	Isolate	6	c	32 ± 12	55 ± 18	+24 ± 20**
II-Human	Casein	17	211 ± 38	207 ± 36	205 ± 35	− 2 ± 10
	Isolate	20	219 ± 42	205 ± 40	197 ± 43	− 8 ± 12****
	Concentrate	20	215 ± 41	199 ± 35	200 ± 38	+ 1 ± 10
II-Rat	Casein	11	96 ± 7	62 ± 7	53 ± 5	− 9 ± 5
	Isolate	12	95 ± 9	63 ± 9	51 ± 5	−13 ± 6*
	Concentrate	12	95 ± 10	63 ± 7	50 ± 6	−13 ± 5**
	Commercial	7	95 ± 10	74 ± 12*****	60 ± 7*****	−14 ± 9
II-Rabbit	Casein	8				
	Isolate	7	45 ± 12	160 ± 54		
	Concentrate	8				
	Commercial	6	45 ± 13	60 ± 17*****	62 ± 26	+ 2 ± 22
II-Rabbit-SP	Casein	12	63 ± 26	121 ± 44	278 ± 119	+157 ± 118
	Isolate	12	54 ± 17	117 ± 48	92 ± 41****	−26 ± 46***
	Concentrate	12	65 ± 38	120 ± 51	95 ± 37****	−26 ± 57***

[a]Diets indicated under Group were given in the test period; for diets given in the control period, see Figure 7-I.

[b]100 mg/dl = 2.59 mmol/l.

[c]Not determined.

Statistical comparison with the casein group by Student's t-test: *$P < 0.075$, **$P < 0.05$, ***$P < 0.001$; with the concentrate group: ****$P < 0.05$; and with all test groups: *****$P < 0.05$.

TABLE 7-IV. The Effect of Casein and Soy Protein Diets on the Concentration of Apolipoproteins in Serum in the Studies With Humans*

	Casein group		Soy isolate group		Soy concentrate group	
	End control period	Change over test period	End control period	Change over test period	End control period	Change over test period
I-Human						
Apolipoprotein-B (mg/l)	488 ± 199	−47 ± 52	462 ± 107	−15 ± 51*		
LDL cholesterol/apo-B^a (g/g)	1.63 ± 0.19	+0.18 ± 0.25	1.92 ± 0.17	−0.08 ± 0.18*		
II-Human						
Apolipoprotein-B (mg/l)	777 ± 202	+67 ± 77	812 ± 201	+12 ± 65* **	783 ± 212	+52 ± 61
LDL cholesterol/apo-B (g/g)	1.66 ± 0.25	−0.10 ± 0.15	1.58 ± 0.19	−0.07 ± 0.21	1.54 ± 0.21	−0.01 ± 0.12
Apolipoprotein-A$_I$ (mg/l)	1,569 ± 242	−32 ± 100	1,552 ± 183	+43 ± 131* **	1,505 ± 276	−29 ± 99
HDL cholesterol/apo-A$_I$^{b} (g/g)	0.35 ± 0.09	0.00 ± 0.02	0.36 ± 0.07	0.00 ± 0.03	0.39 ± 0.06	−0.01 ± 0.03
Apo-A$_I$/Apo-B (g/g)	2.15 ± 0.64	−0.20 ± 0.17	2.01 ± 0.46	+0.04 ± 0.25* **	2.07 ± 0.73	−0.19 ± 0.24

*Results expressed as mean ± SD.

^{a}LDL isolated after ultracentrifugation.

^{b}HDL isolated after Mn-heparin precipitation [Burstein and Samaille, 1960; van der Haar et al, 1978].

Statistical comparison with the casein group by Student's two-tailed t-test: *$P < 0.05$; and with the concentrate group: **$P < 0.05$.

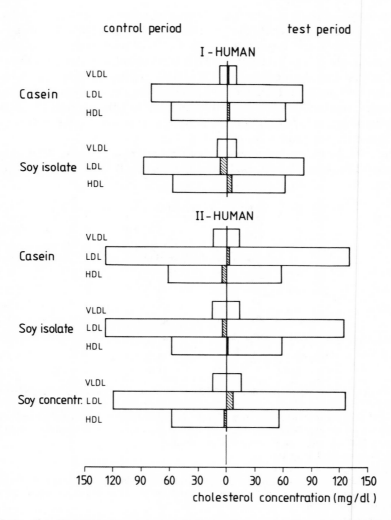

Fig. 7-II.   The effect of casein and soy protein diets on the concentration of cholesterol in lipo-proteins in humans (experiments I-Human and II-Human). The shaded areas right of the vertical line represent an increase over the test period, and those left of the vertical line a decrease; changes less than 1 mg/dl are not shown.

0.86 (n = 56) and r = 0.90 (n = 56), respectively. The correlation coefficients between HDL cholesterol and serum apo-A_I were r = 0.60 (n = 55) and r = 0.67 (n = 55), respectively.

Compared with soy protein concentrate, soy protein isolate produced a decrease in serum total cholesterol, LDL cholesterol, apolipoprotein-B, and LDL cholesterol/apo-B ratio, an increase in HDL cholesterol, and apolipoprotein-A_I, but had no effect on the HDL cholesterol/apo-A_I ratio. These results suggest that the increase in HDL cholesterol has been caused by an increase in the number of HDL particles, and that the decline in LDL cholesterol has been caused by a decline in both the number of LDL particles and in their cholesterol content. No differences in lipoproteins were found between the casein and the soy protein concentrate group.

The overall changes in lipoproteins can be summarized by the HDL cholesterol/total cholesterol ratio and by the apo-A_I/apo-B ratio. In experiment II-

TABLE 7-V. Feed Intake and Weight Gain in Rabbits and Rats

			Feed intake[b]		Weight gain	
Experiment	Groups	Body weight[a] (g)	Control period (g/day)	Test period (g/day)	Control period (g/day)	Test period (g/day)
I-Rabbit[c]	Test (all)[e]	2,322	[f]	75[g]	[f]	12.5
II-Rabbit[d]	Test (all)[e]	1,481	57	57[h,i]	18.2	28.7[i]
	Commercial	1,503	85	85[i]	22.4	25.8[i]
II-Rabbit-SP[d]	Test (all)[e]	1,834	67	67[j]	[f]	17.1
II-Rat[c]	Test (all)[e]	70	9.3	10.6[k]	2.6	1.8
	Commercial	68	11.8	14.0	2.1	1.7

[a]At start control period (II-Rabbit and II-Rat) or at start test period (I-Rabbit and II-Rabbit-SP).
[b]Dry-matter contents of the diets were 94.5% for the pelleted human diets (I-Rabbit and II-Rabbit); 99.5% for the unpelleted human diets (II-Rat); 91.5% for the semipurified diets (II-Rabbit-SP); and 89.0% for the commercial rabbit and rat diets.
[c]Diets were provided ad libitum.
[d]Diets were provided restricted.
[e]Mean data of the test groups are given except for the food intake during the test period in which data for the casein group are given; there were no differences between group means for body weight and weight gain.
[f]Not measured.
[g]Food intake was 83 g per day for the isolate group.
[h]Food intake was 58 and 61 g per day for the isolate and concentrate groups.
[i]During first two weeks of test period.
[j]Food intake was 67.5 and 69.5 g per day for the isolate and concentrate groups.
[k]Food intake was 11.0 and 11.3 g per day for the isolate and concentrate groups.

Human, significant increases in both ratios were observed on the soy isolate diet when compared with the casein or concentrate diet (Table 7-IV; van Raaij et al [1982]).

B. Feed Consumption and Growth of the Experimental Animals

The feed consumption and growth of the rabbits and rats throughout the experiments are presented in Table 7-V. In all of the experiments, except in II-Rabbit, feed consumption and growth were satisfactory. In II-Rabbit, the animals consumed all their feed under the restricted feeding regime throughout the control period and for the first two weeks of the test period. However, after that, twelve rabbits failed to eat all the feed provided, their growth faltered, and their health deteriorated. Four of the rabbits lost hair. It may well be that the human diets, either before or after freeze-drying, contained insufficient vitamins and minerals for the growth of rabbits. When the experiment was completed, the rabbits were transferred to the commercial rabbit diet, and the growth and health of the less severely affected animals improved markedly. Because of the problems encountered, data obtained during the test period of experiment II-Rabbit were deleted in the subsequent analysis of the results. In all other studies, the experimental diets were well accepted, and the feed offered was consumed.

C. Serum Cholesterol and Lipoproteins and Liver Cholesterol in the Experimental Animals

The concentrations of total cholesterol in serum in the experiments with rabbits and rats are shown in Table 7-III. When rabbits were fed semipurified diets (experiment II-Rabbit-SP), the diet containing casein produced a marked increase in serum cholesterol concentration, most of the increased cholesterol being found in the LDL fraction (Table 7-VI). The declines in concentration of cholesterol after the casein control period seen on the two diets containing soy isolate and soy concentrate, respectively, were identical; these declines were observed in the LDL fraction. The higher concentration of cholesterol in the serum on the casein diet compared with the two soy diets is reflected in the concentration of cholesterol in the liver (4.5 ± 1.3 mg cholesterol/g liver, mean $\pm$ SD), although the difference was only significant ($P < 0.05$) when compared with the soy concentrate group (3.3 ± 1.0 mg/g) and not with the soy isolate group (3.6 ± 1.6 mg/g). Most of the additional cholesterol which is found in the liver at increasing serum cholesterol concentrations is in the form of cholesteryl ester. This is seen from the relationships between serum cholesterol concentration and liver cholesteryl ester concentration (Fig. 7-III) and between liver cholesterol concentration and the proportion of cholesterol in the liver that is esterified (Fig. 7-IV).

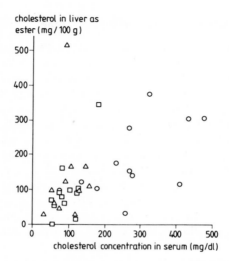

Fig. 7-III.   The relationship between the cholesterol concentration in serum (mg/dl) (x) and the amount of cholesterol as ester in the liver (mg/100 g) (y) in rabbits (experiment II-Rabbit-SP): y = 0.47x + 63.6; r = 0.481, P < 0.01; ○, casein-fed rabbits; △, soy isolate-fed rabbits; □, soy concentrate-fed rabbits.

TABLE 7-VI. The Effect of Casein and Soy Protein Diets on the Concentration of Cholesterol in Lipoproteins in Rabbits on Semipurified Diets (Experiment II-Rabbit-SP)*

	Control period	Test period		
	Casein diet[a]	Casein diet[b]	Soy isolate diet[b]	Soy concentrate diet[b]
Lipoprotein fraction	(n = 36)	(n = 12)	(n = 12)	(n = 12)
VLDL	17	33	17	18
LDL	74	223	48	55
HDL	25	41	31	31
Whole serum[c]	119	278	92	95

*All values expressed in mg/100 ml.
[a]Mean values of six pools, each of serum from 6 animals.
[b]Mean values of four pools, each of serum from 3 animals.
[c]Mean values from serum of individual animals.

In the first experiment with rabbits using the human diets (experiment I-Rabbit), the rabbits were transferred straight from the commercial diet to the test diets containing either casein or soy protein isolate. After 2½ weeks of test period, the casein diet produced significantly higher increases in cholesterol than did the soy isolate diet (Table 7-III). When the rabbits were subsequently returned to the commercial diet, the concentration of cholesterol in the serum of both groups of rabbits fell to a mean value of about 30 mg/dl [van Raaij et al, 1981]. Thus the human diet containing casein has a pronounced hypercholesterolemic effect in rabbits when compared with the soy protein diet.

In the second experiment with rabbits using the human diets (experiment II-Rabbit), the rabbits were transferred from the commercial diet to the human diets for a control period of 2½ weeks before being transferred to the test diets for a subsequent four weeks. On the casein control diet, the concentrations of cholesterol in serum increased steeply and significantly when compared with the cholesterol levels of the rabbits that continued on the commercial diet. After 4–5 weeks on the human diets, the rabbits began to show signs of ill health such as reduced appetite, reduced growth and hair loss. As discussed earlier, this may have been due to a mineral or vitamin deficiency, which did

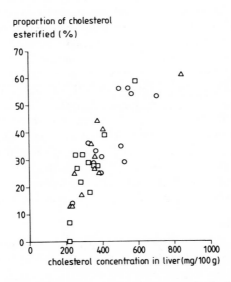

Fig. 7-IV. The relationship between the cholesterol concentration in the liver (mg/100 g) (x) and the proportion of cholesterol esterified (%) (y) in rabbits (experiment II-Rabbit-SP): y = 0.087x − 1.46; r = 0.827, P < 0.001; ○, casein-fed rabbits; △, soy isolate-fed rabbits; □, soy concentrate-fed rabbits.

not become manifest within the 2½-week period that the rabbits in experiment I-Rabbit were on the human diets.

In the experiment with rats using the human diets (II-Rat), the rats were transferred from the commercial diet to the human casein diet for a control period of 2½ weeks before being transferred to the human test diets. Concentration of cholesterol in serum was declining in the rats used, as shown by the rats that continued on the commercial diet for the whole experiment. On the casein control diets significant lower concentrations of cholesterol in serum were observed than on the commercial diet. During the test period the declines in serum cholesterol on the human test diets were similar to those on the commercial diet; however, the declines on both soy protein diets were more pronounced than that on the casein diet (Table 7-III). The larger declines on the soy protein diets occurred mainly in the lipoprotein fractions with densities between 1.060 and 1.128 g/ml (Table 7-VII).

IV. DISCUSSION

In our studies with young and middle-aged adults we failed to find a marked effect on serum total cholesterol of diets containing either soy isolate or soy concentrate when compared with a diet containing casein. When the soy preparations were incorporated into semipurified diets and fed to rabbits, we obtained results similar to those reported previously (for a review, see Terpstra et al [1982b]): Casein was found to be strongly hypercholesterolemic when compared with the soy preparations. Similar results were observed when rabbits were fed the human diets: A human diet containing soy isolate resulted in significantly lower cholesterol concentrations than the human diet containing casein (experiment I-Rabbit). In experiment II-Rabbit, the rabbits also showed a marked increase in serum cholesterol concentration after they had been transferred from the commercial diet to the casein-containing human diet (Table 7-III). As discussed before, it is unfortunate that in experiment II-Rabbit no comparisons could be made between the casein and the soy protein diets. In rats, the differences between the human diets containing either casein or soy protein were small, but marginally significant.

How can one explain these different results between humans and animals?

A. Species Effects

Animal species differ greatly in their resistance to diet-induced hypercholesterolemia. For example, rabbits are highly susceptible to hypercholesterolemia by dietary means; the replacement of protein in semipurified diets by casein and the addition of small amounts of cholesterol have both been shown to be very effective. Although most studies with rabbits have been carried out with semipurified diets, our studies clearly indicate that human diets give similar re-

TABLE 7-VII. The Effect of Casein and Soy Protein Diets on the Concentration of Cholesterol in Lipoproteins in Rats (Experiment II-Rat)*

Lipoprotein fraction (density [d] in g/ml)	Casein group			Soy isolate group			Soy concentrate group		
	Control period	Test period	Change	Control period	Test period	Change	Control period	Test period	Change
d < 1.016	2.9	2.5	− 0.4	2.5	2.1	− 0.4	2.9	2.1	− 0.8
1.016 < d < 1.032	1.5	1.5	0.0	1.4	1.4	0.0	1.4	1.2	− 0.2
1.032 < d < 1.045	1.7	2.1	+ 0.4	1.9	2.1	+ 0.2	1.9	1.9	0.0
1.045 < d < 1.060	3.5	3.5	0.0	4.6	4.1	− 0.6	3.5	3.9	+ 0.4
1.060 < d < 1.078	8.1	7.7	− 0.4	10.6	9.7	− 1.0	8.3	6.6	− 1.7
1.078 < d < 1.102	22.4	18.6	− 3.9	23.4	15.3	− 8.1	24.6	20.5	− 4.1
1.102 < d < 1.128	13.2	12.0	− 1.2	14.1	10.1	− 4.1	13.5	11.4	− 2.1
1.128 < d < 1.146	2.5	2.9	+ 0.4	2.7	3.3	+ 0.6	2.9	3.3	+ 0.4
1.146 < d	2.1	3.5	+ 1.4	1.9	3.7	+ 1.7	2.1	2.9	+ 0.8
Whole serum (pools)	66.0	52.6	−13.4	69.1	51.5	−17.6	65.8	51.7	−14.1

*Results expressed in mg/100 ml of serum. At the end of the control and test period, two serum samples were taken at one-day intervals. Serum samples were pooled per group. Each given value represents the average value of the two pools at the end of the respective period.

sults. In our studies the effects with the human diets were less pronounced than with the semipurified diets, but this can be explained, at least partly, by the fact that in the human diets only 65% of the protein consisted of test protein compared with 100% in the semipurified diets, while the total proportion of protein in the two types of diets was about the same. The hypercholesterolemic effect of casein increases when the amount of casein in the diet is increased [Terpstra et al, 1981a]. The amount of linoleic acid in the diet may also be a contributing factor as the casein-induced hypercholesterolemia in rabbits disappears when increased amounts of linoleic acid are included in the diets [Lambert et al, 1958; Wigand, 1959; Carroll and Hamilton, 1975; Beynen and West, 1981]. In our human diets, the proportion of polyunsaturated fat in the diet was about 4% by weight, while in the semipurified diets the proportion was less than 1%. The differences in results between the human and semipurified diets can probably not be explained by difference in duration, because after 2½ weeks on the semipurified diets the difference in effects between the casein and soy protein diets was already greater than on the human diets (data not shown).

In rats fed semipurified diets the effects of dietary proteins are only observed when considerable amounts of cholesterol (approximately 1%) are added to the diets [Terpstra et al, 1982a]. In our experiments only 0.07% of cholesterol was present, and this explains why the observed effects in rats were so small. Our studies with humans suggest that normocholesterolemic adults are also relatively insensitive to changes in dietary protein, and one might speculate that in humans, diets have to contain considerable amounts of cholesterol before the differential effect of casein and soy protein on the level of serum cholesterol is expressed. It should be noted, however, that in the experiments of Sirtori et al [1979] the cholesterol-lowering effect of the soy diet in hypercholesterolemic patients was most pronounced against a high-linoleic acid low-cholesterol background, ie, the opposite of the conditions under which the effects of dietary proteins are seen most clearly in animals.

B. Age of the Animals

Animal studies in general and our studies in particular have been carried out with young, growing animals, while in most human studies, including those reported here, adults were used. It has been reported that dietary casein-induced hypercholesterolemia in male rabbits occur only in young growing animals [West et al, 1982], and similar results have been observed in rats [McGregor et al, 1971]. Such animals spend a considerable proportion of their lives on the experimental diets. Therefore, our results on humans do not rule out the possibility that a long-term intake of animal protein starting at an early age will cause similar cholesterol-raising effects in normocholesterolemic humans.

C. Differences in Lipoprotein Metabolism

A point to keep in mind in extrapolating animal data to man concerns the differences in the distribution of cholesterol over the various lipoprotein fractions. In healthy humans, most of the serum cholesterol is transported in the LDL fraction, whereas in normocholesterolemic rabbits and rats the HDL fraction is the main carrier of cholesterol [Terpstra et al, 1982c]. Furthermore, hypercholesterolemic diets in animals often cause the appearance of an unusual lipoprotein with the density of VLDL but with a much higher cholesterol content and with a higher proportion of apolipoprotein-E [Ross and Zilversmit, 1977; Mahley et al, 1980; Terpstra et al, 1982a; Scholz et al, 1982]. In humans such a particle is seen only in type III hypercholesterolemia, a fairly rare disease. In healthy humans we did not find any marked abnormalities of the lipoprotein spectrum after six weeks on the casein diet. In comparison, however, the subjects who had received the soy isolate diet showed slightly lower LDL and higher HDL cholesterol levels. Apart from slight changes in the composition of LDL in I-Human, our apoprotein measurements did not point to marked effects of dietary proteins on lipoprotein composition in man. Thus at the level of the separate lipoproteins, casein-fed rabbits and rats appear to be different from humans.

D. Possible Mechanisms

The mechanisms underlying the cholesterolemic properties of dietary protein are not clearly understood. Probably several mechanisms are involved. The amino acid composition of the proteins plays a role [Carroll, 1978; Kritchevsky, 1979; Hermus, 1975; Terpstra et al, 1982c], but how amino acids might influence cholesterol metabolism remains to be established. Differences in amino acid composition between casein and soy protein, however, cannot fully explain the different results [Carroll, 1978; Terpstra et al, 1982c]. It has been suggested that some substances in the nonprotein part of the soy preparations such as fiber or saponins might be partly responsible for the observed effects [Potter et al, 1980; Topping et al, 1980], but our human and animal studies clearly indicate that the nonprotein part of soy concentrate did not influence serum cholesterol when compared with soy isolate. Similar results were found by Hamilton and Carroll [1976] in rabbits. As discussed previously [van Raaij et al, 1982] the type of dietary fiber present in soy concentrate is not likely to exert a marked hypocholesterolemic effect. As casein increases the concentration of cholesterol in both the liver and in serum (experiment II-Rabbit-SP), a redistribution of cholesterol between the serum and liver would not appear to be involved as is the case with the difference between rabbits which are hypo- or hyperresponsive to cholesterol [West and Roberts, 1974]. Perhaps the differential effect of casein and soy protein on serum cholesterol levels in rabbits may be attributible to differences in the digestibility of the two proteins. As has been postulated elsewhere [Terpstra et al, 1982c] soy protein may be less digestible than casein in rabbits and the undigested protein could bind bile acids, thus facilitating their excretion and preventing their reabsorption. It may well be that in humans, soy protein is not significantly less digestible than casein or that the binding of soy protein to bile acids does not have a significant effect on the excretion of bile acids either because of the presence of other materials in the gut or because of the nature of the bile acids involved.

Although we did not find a marked change in serum total cholesterol concentrations in humans on the casein and soy diets, our data do show a small decline in LDL cholesterol and a small increase in HDL cholesterol on the soy isolate diet when compared with the casein diet, but there was no effect with the soy concentrate diet. Apart from small changes in the composition of LDL in I-Human, the apoprotein results suggest that the changes in cholesterol concentrations in the lipoprotein fractions resulted from changes in the number of lipoprotein particles.

An increase in the ratio of HDL cholesterol/total cholesterol has been associated with a lower risk of coronary heart disease [Blackburn et al, 1977; Miller and Miller, 1975], so it may well be that the increase in this ratio on the soy isolate diet could have a beneficial effect even when the concentration of total cholesterol remains constant. Yet it must be noted that in our studies these favorable effects were only observed with the rather pure soy protein isolate and not with the soy concentrate. Thus the dramatic effects observed by Sirtori et al [1979] with textured soy protein not only on HDL cholesterol/total cholesterol ratio but also on the concentration of total cholesterol eluded us when we used the soy concentrate diet.

An expectation has developed that the replacement of animal proteins by vegetable proteins in human diets might aid in lowering serum total cholesterol levels, thereby providing a very useful tool in the prevention of coronary heart disease [Lewis, 1980]. This expectation is based on the many studies with animals, particularly the rabbit, and a limited number of studies with humans, particularly hypercholesterolemic patients [Sirtori et al, 1979; Descovich et al, 1980; Wolfe et al, 1981] in which diets containing animal proteins, particularly casein, have been compared with diets containing vegetable proteins, particularly soy protein. However, the results of most studies with normocholesterolemic humans, including our two studies, do not provide support for this expectation: The normocholesterolemic human appears to be relatively insensitive to changes in the type of protein in the diet. However, a small favorable effect on the distribution of cholesterol over the various lipoprotein classes cannot be excluded.

V. SUMMARY

This paper reports the results of studies in which the effect of casein and soy protein on serum cholesterol and lipoprotein concentrations were compared in humans, rabbits, and rats using human diets. In the human studies, no marked effect of diets containing either soy isolate or soy concentrate on serum total cholesterol was observed when compared with a diet containing casein. When rabbits and rats were fed the human diets, lower serum cholesterol levels were found on both the soy diets when compared with the casein diet, the differences being much more pronounced in rabbits than in rats. These results confirm differences in susceptibility between species, and that the normocholesterolemic human appears to be relatively insensitive to changes in these dietary proteins. Although in humans no effects on serum total cholesterol level were observed, the soy isolate diet did cause a small decline in cholesterol concentration in the low-density lipoprotein fraction (LDL) and a small increase in cholesterol concentration in the high-density lipoprotein fraction (HDL) when compared with the casein diet. However, there was no effect with the soy concentrate diet. Analysis of apolipoprotein concentrations suggested that the changes in cholesterol concentrations in the lipoprotein fractions resulted mainly from changes in the number of lipoprotein particles, but minor effects on the composition of the LDL could not be excluded. The lack of a cholesterol-lowering effect of the less-refined soy concentrate when compared with soy protein isolate in both the human and animal studies suggest that the non-protein part of the soy preparations probably does not have a specific cholesterol-lowering effect. Our studies stress the risk of extrapolating animal data concerning the effect of protein on cholesterol metabolism to man.

ACKNOWLEDGMENTS

This work was supported in part by grants from the Netherlands Institute for Dairy Research (NIZO) and The Netherlands Heart Foundation (79.045, 26.003).

The authors are most grateful to the volunteers for their invaluable cooperation. We are indebted to the firms involved in the development and preparation of the special food products for the human studies (see van Raaij et al [1981, 1982]); to J.B. Schutte and K. Deuring (Institute for Animal Nutrition Research ILOB-TNO, Wageningen) for preparing the animal diets and performing the rabbit experiments; to G. van Tintelen and J. Haas for performing the rat experiment; to H. Koopman for freeze-drying the human diets; to P.H.E. Groot (Department of Biochemistry I, Erasmus University, Rotterdam) and P.N.M. Demacker (Department of Internal Medicine, University of Nijmegen) for providing us with the methodology and materials for apolipoprotein assays; to the technicians Mrs J.H.M. Barendse-van Leeuwen, Mrs E.A.M. Koot-Gronsveld, Miss A.E.M.F. Soffers, F.J.M. Schouten, and H.J. Verwey; and to A.H.M. Terpstra for helpful advice and discussion.

VI. REFERENCES

Anderson JT, Grande F, Keys A (1971): Effect on man's serum lipids of two proteins with different amino acid composition. Am J Clin Nutr 24:524–530.

Beynen AC, West CE (1981): The distribution of cholesterol between lipoprotein fractions of serum from rabbits fed semipurified diets containing casein and either coconut oil or corn oil. Z Tierphysiol Tierernaeher Futtermittelk 46:233–239.

Blackburn H, Chapman J, Dawber TR, Doyle JT, Epstein FH, Kannel WB, Keys A, Moore F, Paul O, Stamler J, Taylor HL (1977): Revised data for 1970 ICHD report. Am Heart J 94: 539–540.

Bodwell CE, Schuster EM, Steele PS, Judd JT, Smith JC (1980): Effects of dietary soy protein on plasma lipid profiles of adult men. (Abstr 4456), Fed Proc 39:1113.

Brussaard JH, Dallinga-Thie GM, Groot PHE, Katan MB (1980): Effect of amount and type of dietary fat on serum lipids, lipoproteins and apoproteins in man: A controlled 8-week trial. Atherosclerosis 36:515–527.

Burslem J, Schonfeld G, Howald MA, Weidman SW, Miller JP (1978): Plasma apoprotein and lipoprotein lipid levels in vegetarians. Metabolism 27:711–719.

Burstein M, Samaille J (1960): Sur un dosage rapide du cholesterol lié au α- et β-lipoproteins du serum. Clin Chim Acta 5:609.

Carroll KK (1978): The role of dietary protein in hypercholesterolemia and atherosclerosis. Lipids 13:360–365.

Carroll KK, Hamilton RMG (1975): Effects of dietary protein and carbohydrate on plasma cholesterol levels in relation to atherosclerosis. J Food Sci 40:18–23.

Carroll KK, Giovannetti PM, Huff MW, Moase O, Roberts DCK, Wolfe BM (1978): Hypocholesterolemic effect of substituting soybean protein for animal protein in the diet of healthy young women. Am J Clin Nutr 31:1312–1321.

Descovich GC, Gaddi A, Mannino G, Cattin L, Senin U, Caruzzo C, Fragiocomo C, Sirtori M, Ceredi C, Benassi MS, Colombo L, Fontana G, Mannarino E, Bertelli E, Noseda G, Sirtori CR (1980): Multicentre study of soybean protein diet for outpatient hypercholesterolaemic patients. Lancet ii:709–712.

Folch J, Lees M, Stanley GHS (1957): A simple method for the isolation and purification of total lipids from animal tissues. J Biol Chem 226:497.

Haar F van der, Gent CM van, Schouten FM, Voort HA van der (1978): Methods for the estimation of high density cholesterol, comparison between two laboratories. Clin Chim Acta 88: 469–481.

Hamilton RMG, Carroll KK (1976): Plasma cholesterol levels in rabbits fed low fat, low cholesterol diets. Effects of dietary proteins, carbohydrates and fibre from different sources. Atherosclerosis 24:47–62.

Hardinge MG, Stare FJ (1954): Nutritional studies of vegetarians. 2. Dietary and serum levels of cholesterol. Am J Clin Nutr 2:83–88.

Hermus RJJ (1975): "Experimental Atherosclerosis in Rabbits on Diets With Milk Fat and Different Proteins." Wageningen, The Netherlands: Centre for Agricultural Publishing and Documentation.

Holmes WL, Rubel GB, Hood SS (1980): Comparison of the effect of dietary meat versus dietary soybean protein on plasma lipids of hyperlipidemic individuals. Atherosclerosis 36: 379–387.

Huang TC, Chen CP, Wefler V, Raftery A (1961): A stable reagent for the Liebermann-Burchard reaction. Application to rapid serum cholesterol determination. Anal Chem 33: 1405-1407.

Katan MB, Vroomen LHM, Hermus RJJ (1982): Reduction of casein-induced hypercholesterolemia and atherosclerosis in rabbits and rats by dietary glycine, arginine and alanine. Atherosclerosis 43:381-391.

Kritchevsky D (1979): Vegetable protein and atherosclerosis. J Am Oil Chem Soc 56:135-140.

Kritchevsky D, Tepper SA, Williams DE, Story JA (1977): Experimental atherosclerosis in rabbits fed cholesterol-free diets. 7. Interaction of animal or vegetable protein with fiber. Atherosclerosis 26:397-403.

Lambert GF, Miller JP, Olsen RT, Frost DV (1958): Hypercholesterolemia and atherosclerosis induced in rabbits by purified high fat rations devoid of cholesterol. Proc Soc Exp Biol Med 97:544-549.

Laurell CB (1972): Electroimmunoassay. Scand J Clin Lab Invest (Suppl) 124:21.

Lewis B (1980): Dietary prevention of ischaemic heart disease — A policy for the '80s. Br Med J 281:177-180.

Mahley RW, Innerarity TL, Brown MS, Ho WK, Goldstein JL (1980): Cholesteryl ester synthesis in macrophages: Stimulation by β-very low density lipoproteins from cholesterol-fed animals of several species. J Lipid Res 21:970.

McGregor D (1971): The effects of some dietary changes upon the concentrations of serum lipids in rats. Br J Nutr 25:213-224.

Miller GJ, Miller NE (1975): Plasma high-density lipoprotein concentration and development of ischaemic heart-disease. Lancet i:16-19.

Potter JD, Illman RJ, Calvert GD, Oakenfull DG, Topping DL (1980): Soya saponins, plasma lipids, lipoproteins and fecal bile acids: A double blind cross-over study. Nutr Rep Int 22: 521-528.

Raaij JMA van, Katan MB, Hautvast JGAJ, Hermus RJJ (1981): Effects of casein versus soy protein diets on serum cholesterol and lipoproteins in young healthy volunteers. Am J Clin Nutr 34:1261-1271.

Raaij JMA van, Katan MB, West CE, Hautvast JGAJ (1982): Influence of diets containing casein, soy isolate and soy concentrate on serum cholesterol and lipoproteins in middle-aged volunteers. Am J Clin Nutr 35:925-934.

Redgrave TG, Roberts DCK, West CE (1975): Separation of plasma lipoproteins by density gradient ultracentrifugation. Anal Biochem 65:42-49.

Röschlau P van, Bernt E, Gruber W (1974): Enzymatische Bestimmung des Gesamt-Cholesterins im Serum. Z Klin Chem Klin Biochem 12:403-407.

Ross AC, Zilversmit DC (1977): Chylomicron remnant cholesteryl esters as the major constituent of very low density lipoproteins in plasma of cholesterol-fed rabbits. J Lipid Res 18:169.

Sacks FM, Castelli WP, Donner A, Kass EH (1975): Plasma lipids and lipoproteins in vegetarians and controls. N Engl J Med 292:1148-1151.

Scholz KE, Beynen AC, West CE (1982): Comparison between the hypercholesterolaemia in rabbits induced by semipurified diets containing either cholesterol or casein. Atherosclerosis 44:85-97.

Shorey RL, Davis JL (1979): Effects of substituting soy for animal protein in the diets of young mildly hypercholesterolemic males. (Abstr), Fed Proc 38:551.

Sirtori CR, Gatti E, Montero O, Conti F, Agradi E, Tremoli E, Sirtori M, Fraterrigo L, Tavazzi L, Kritchevsky D (1979): Clinical experience with the soybean protein diet in the treatment of hypercholesterolemia. Am J Clin Nutr 32:1645-1658.

Snedecor GW, Cochran WG (1967): "Statistical Methods." Ames: Iowa State University Press.

Stamler J (1979): Population studies. In Levy RI, Rifkind BM, Dennis BH, Ernst ND (eds): "Nutrition, Lipids and Coronary Heart Disease." New York: Raven Press, pp 25-88.

Terpstra AHM, Sanchez-Muniz FJ (1981): Time course of the development of hypercholesterolemia in rabbits fed semipurified diets containing casein or soybean protein. Atherosclerosis
39:217–227.

Terpstra AHM, Harkes L, van der Veen FH (1981a): The effect of different proportions of
casein in semi-purified diets on the concentration of serum cholesterol and the lipoprotein
composition in rabbits. Lipids 16:114–119.

Terpstra AHM, Woodward CJH, Sanchez-Muniz FJ (1981b): Improved techniques for the
separation of serum lipoproteins by density gradient ultracentrifugation: Visualization by
prestaining and rapid separation of serum lipoproteins from small volumes of serum. Anal
Biochem 111:149–157.

Terpstra AHM, van Tintelen G, West CE (1982a): The effects of semipurified diets containing
different proportions of either casein or soybean protein on the concentration of cholesterol
in whole serum, serum lipoproteins and liver in male and female rats. Atherosclerosis 42:
85–95.

Terpstra AHM, Hermus RJJ, West CE (1982b): The role of dietary protein in cholesterol metabolism. World Rev Nutr Diet (in press).

Terpstra AHM, Hermus RJJ, West CE (1982c): Dietary protein and cholesterol metabolism in
rabbits and rats. In Kritchevsky D, Gibney MJ (eds): "Current Topics in Nutrition and Disease: Animal and Vegetable Proteins in Lipid Metabolism." New York: Alan R. Liss Inc,
pp 19–49.

Topping DL, Trimble RP, Illman RJ, Potter JD, Oakenfull DG (1980): Prevention of dietary
hypercholesterolemia in the rat by soy flour high and low in saponins. Nutr Rep Int 22:513–
519.

Vessby B, Boberg J, Gustafsson IB, Karlström B, Lithell H, Werner I (1980): Substitution with
textured soy protein for animal protein in a lipid lowering diet — Effects on the serum lipoprotein composition. (Abstr), Eur J Clin Invest 11:1823.

West CE, Roberts DCK (1974): Cholesterol metabolism in two strains of rabbits differing in
their cholesterolaemic response to dietary cholesterol. Biochem Soc Trans 2:1275–1277.

West CE, Rowbotham TR (1967): The use of a computer in the determination by gas-liquid
chromatography of the concentrations and identification of individual fatty acids present as
free fatty acids, triglycerides and cholesteryl esters. J Chromatogr 30:62–76.

West CE, Deuring K, Schutte JB, Terpstra AHM (1982): The effect of age on the development
of hypercholesterolemia in rabbits fed semipurified diets containing casein. J Nutr 112:1287–
1295.

Wigand G (1959): Production of hypercholesterolemia and atherosclerosis in rabbits by feeding
different fats without supplementary cholesterol. Acta Med Scand 166 (Suppl 351):1–91.

Wolfe BM, Giovannetti PM, Cheng DCH, Roberts DCK, Carroll KK (1981): Hypolipidemic
effect of substituting soybean protein isolate for all meat and dairy protein in the diets of hypercholesterolemic men. Nutr Rep Int 24:1187–1198.

Animal and Vegetable Proteins in Lipid
Metabolism and Atherosclerosis, pages 135–148
© *1983 Alan R. Liss, Inc., 150 Fifth Ave., New York, NY 10011*

8
Studies on the Use of a Soybean Protein Diet for the Management of Human Hyperlipoproteinemias

C.R. Sirtori, G. Noseda, and G.C. Descovich

Center E. Grossi Paoletti, University of Milano, Italy (C.R.S.), Beata Vergine Hospital, Mendrisio, Switzerland (G.N.), and II Medical Clinic, University of Bologna, Italy (G.C.D.)

I. INTRODUCTION

Early attempts to substitute animal proteins in a diet with textured vegetable proteins (TVP) from soybean, started in 1972 at the Center for the Study of Hyperlipidemias of the University of Milano. The only objective, at that time, was to find a dietary substitute which would allow raising the polyunsaturated/saturated (P/S) fatty acid ratio beyond the limits usually reached by diets with animal proteins. The considerable backlog of knowledge concerning the postulated hypolipidemic and antiatherosclerotic properties of vegetable proteins [Sirtori, 1982] was unknown to the investigators. Interestingly, a clinical study with a similar protocol had been carried out a few years before [Hodges et al, 1967], also with different objectives, and with results not different from those later reported by us.

These 10-year-old studies, carried out in hypercholesterolemic volunteers admitted to a metabolic ward, clearly showed that the simple substitution of animal proteins with TVP resulted in a significant decrease of plasma total and low-density lipoprotein (LDL) cholesterol levels. This reduction far exceeded that predictable from variations of only the cholesterol content or of the P/S ratio of the diets under study [Keys et al, 1965]. These findings prompted a wide extention of the clinical investigations, particularly in individuals with type II hyperlipidemia, involving, up to now, many hundreds of patients within several research protocols.

This presentation will review our clinical experience with this dietary treatment, mostly in type II hyperlipoproteinemic patients. Clinical work, pertaining to the mechanism of action of the soybean diet, will also be analyzed, with particular interest for cholesterol balance studies and for changes in plasma amino acid and hormonal profiles.

II. CLINICAL EVALUATION OF THE SOYBEAN PROTEIN DIET TREATMENT

A. Short-term Comparative Studies

After the early attempts to administer a low-lipid, low-cholesterol diet with total substitution of animal proteins with TVP (unpublished observations), 20 type II patients participated in a crossover trial, involving three weeks of administration of a standard hypolipidemic regimen (cholesterol < 300 mg, P/S 2.0) and three weeks of the TVP diet, with a very similar lipid composition [Sirtori et al, 1977]. This trial, although hampered by a limited experience with adequate recipes for the use of TVP, still indicated remarkable differences between the effects of a standard hypolipidemic diet (previously already used by all patients) and those of the soybean diet (Fig. 8-I). In almost all cases, the TVP diet induced a plasma cholesterol reduction of 15% or more as compared to the standard regimen. On the whole, the difference between the plasma cholesterol levels at the end of the two dietary treatments was more than 20%, almost all being explained by changes in LDL cholesterol.

In a small group of type II patients, also admitted to the metabolic ward, the soybean diet treatment was given for six weeks. During three of these, either the first or the second, 500 mg of cholesterol was added daily. The results of this second experiment were clearly indicative of a cholesterol-lowering effect of the soybean diet, independent of the dietary cholesterol intake, thus ruling out any direct contribution of the very low cholesterol content of the experimental diet [Sirtori et al, 1977].

In other protocols, different factors were evaluated, always within short-term (usually six weeks) metabolic ward studies. The significance of the P/S ratio in the diet was tested in a crossover trial, comparing three weeks of soybean diet with a low P/S (0.1) and three weeks with a high P/S (2.7), given in this, or in the reverse sequence [Sirtori et al, 1979]. These studies clearly showed that switching from a low to a high P/S ratio, or vice versa, was associated with a plasma cholesterol change, predictable from the Keys' formula

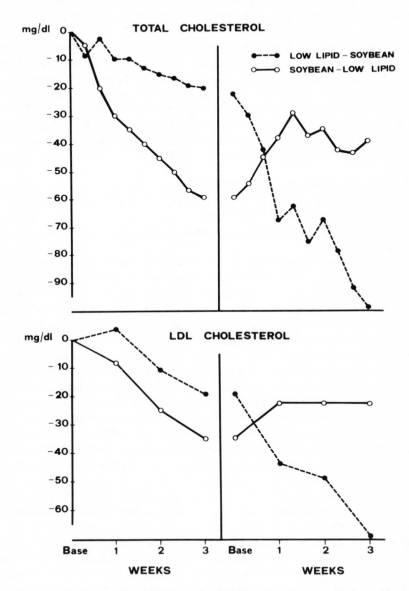

Fig. 8-I. Results of a crossover trial comparing, in 20 type II patients, the cholesterol-lowering activity of the soybean diet and of a standard low lipid diet with animal proteins [Sirtori et al, 1977].

[Keys et al, 1965]. On the other hand, the soybean-diet treatment elicited a plasma cholesterol reduction, always exceeding 15%, independent from the P/S changes.

The cholesterol-lowering properties also appeared not to be associated with a reduced intake of sulfur containing amino acids. Methionine administration, in a daily amount adequate to equalize the amino acid pattern of egg albumin [Helms et al, 1977], failed to modify the pattern of cholesterol reduction.

Some of the patients participating in these short-term studies were followed for more prolonged periods at home, while following a similar dietary regimen. In about half of them, the plasma cholesterol reduction achieved in the hospital could be either maintained or further improved [Sirtori et al, 1979].

B. Longer-term Treatments in Outpatients

The considerable experience, acquired during the inpatient studies with the soybean regimen, as well as the availability of new preparations more palatable for Italian patients suggested that similar studies be also carried out in outpatient type II subjects.

A large study, involving 130 patients with plasma cholesterol exceeding 300 mg/dl, was carried out in several lipid clinics in Italy and Switzerland. The participating patients were followed for four weeks while on their habitual low-lipid, low-cholesterol regimen, after which they were given the soybean diet for a period of eight weeks. At the end of this, the patients were followed for a further six weeks on their habitual diet. One hundred twenty-seven patients completed the study, thus allowing us to conclude that the regimen induces a plasma cholesterol reduction of the same degree as observed during the metabolic ward studies [Descovich et al, 1980]. Plasma lipoprotein fractionation also indicated that the decrease in cholesterolemia was totally attributable to a drop in LDL levels; high-density lipoprotein (HDL) cholesterol levels were not modified. Follow-up of the patients after the end of the soybean-diet treatment resulted in a slow but progressive reincrease of cholesterolemia. This trend was more marked in patients with a familial form of the disease (Fig. 8-II).

Similar findings have been reported in other studies in Western Europe, when the soybean diet was substituted totally [Liebermeister and Toluipur, 1980] or partially [Karlström et al, 1979; Schwandt et al, 1981] in the standard hypolipidemic regimen, in patients with clear-cut type II hyperlipoproteinemia. However, contrasting findings (ie, negligible plasma cholesterol reductions) have been reported by other authors. In these studies, however, the participating subjects were mostly those with borderline plasma cholesterol elevations [Holmes et al, 1980; Shorey et al, 1981] and/or essentially normolipidemic [van Raaij et al, 1981]. Significant differences, moreover, may exist be-

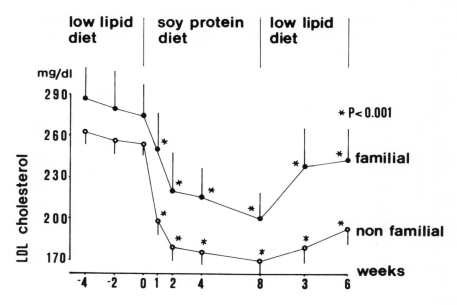

Fig. 8-II.   Changes in low-density lipoprotein (LDL) cholesterol levels during an outpatient trial with the soy protein diet in type II patients. Both familial (more than three affected relatives) and nonfamilial cases show a significant hypocholesterolemic response; the former group displays, however, a faster return to elevated cholesterol levels [Descovich et al, 1980].

tween the types of diet used in the reported studies and those selected by the quoted authors [Sirtori et al, 1981].

The overall analysis of the plasma cholesterol reductions recorded in all of our formal protocols (be they on in- or outpatients) are summarized in Figure 8-III. A statistically significant negative correlation between the plasma cholesterol drop and the starting cholesterolemia is clearly evident. This figure also indicates that only minor plasma cholesterol variations are to be expected in subjects with normal cholesterolemia, or with borderline cholesterol elevations.

The availability of a palatable preparation has also allowed a long-term evaluation of the maintenance of the hypocholesterolemic effect in selected, well-motivated patients. Up to now, 27 patients (13 males, 14 females, age range 20–60 years) have been followed, while on this regimen, for over two years. After at least two months of total substitution of the dietary proteins, the patients were instructed to take at least six meals per week with the textured

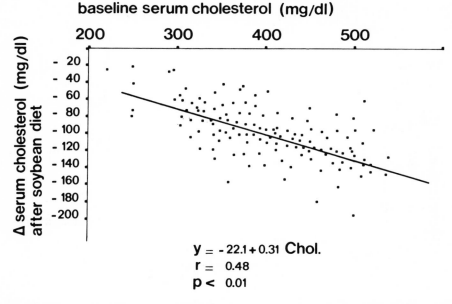

Fig. 8-III. Correlation between baseline serum cholesterol levels, at the beginning of each formal dietary trial, and of the observed reduction of cholesterolemia at the end of the same. The figures are taken from all our published studies up to 1981.

product. No significant side effects have been encountered, except for the difficult compliance with a relatively demanding dietary treatment.

Altogether, the results are consistent with the maintenance of the hypocholesterolemic activity, up to two years of treatment and longer. In these patients, a small but progressive increase in HDL cholesterol levels may also be noted (Fig. 8-IV). The clinical evaluation of some of the treated subjects also indicates that possibly atherosclerotic changes may regress with a continued dietary treatment. As shown in Figure 8-V, ECG changes, consistent with a severe subendocardial ischemia in a 32-year-old woman, regressed, without any drug treatment, after three years of treatment with the soybean diet, eliciting a plasma cholesterol reduction from 332 to 206 mg/dl with normal triglyc-

Fig. 8-IV. Long-term follow-up of total and HDL cholesterol levels in 27 patients with type II hyperlipoproteinemia, on the regimen for over 18 months. The plasma cholesterol reduction is maintained while on six meals per week of soybean diet. A significant HDL cholesterol rise is noted, particularly in females, at the long term intervals.

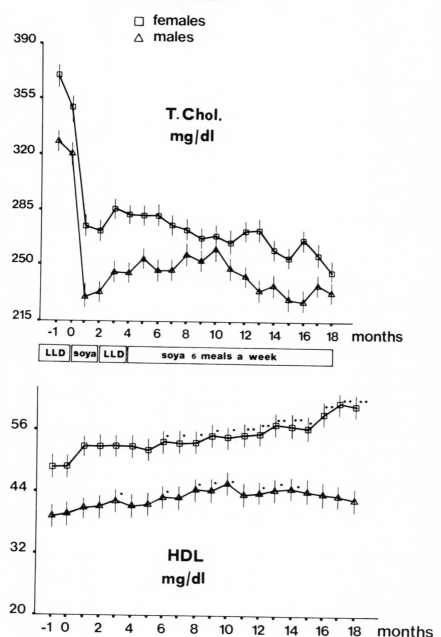

SOYA PROTEIN (CHOLSOY)
LONG TERM DIET

□ females
△ males

T. Chol.
mg/dl

| LLD | soya | LLD | soya 6 meals a week |

HDL
mg/dl

BEFORE   1s   2s

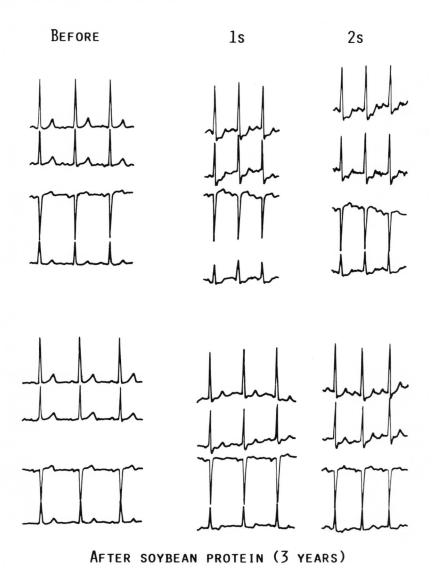

AFTER SOYBEAN PROTEIN (3 YEARS)

Fig. 8-V.   Electrocardiographic evaluation of a 32-year-old female with type II hyperlipoproteinemia before (upper tracings) and after three years of soybean diet administration (lower
tracings). No changes are noted at rest (left-hand tracings). Both at the first and second step of
the Bruce protocol (1S-2S) a significant ST-depression is noted both in the standard and pre-
cordial leads. These alterations are not detectable after 3 years of diet.

erides (68 to 59 mg/dl). In the same subject, a Thallium scintigram of the ventricles taken at the two-year follow-up gave normal findings.

III. CLINICAL STUDIES ON THE MECHANISM OF ACTION OF THE SOYBEAN DIET TREATMENT

A. Sterol Biosynthesis and Excretion in Humans Following Different Dietary Proteins

A limited number of studies have described changes in the biosynthesis and excretion of sterols in humans following dietary protein changes. An early report by Potter and Nestel [1976] in infants, carried out with different objectives, described a significant increase of neutral sterol excretion, upon shifting from standard milk to soybean milk. More recently, the same authors reported a reduced LDL biosynthesis in vegetarians, together with a low excretion of neutral sterols and bile acids [Nestel et al, 1981]. These findings, at variance with the expected changes in sterol excretion in the presence of a high dietary-fiber intake [Miettinen and Tarpila, 1977], are consistent with a mode of action unrelated to nonabsorbable dietary components. In vegetarians, Huijbregts et al [1980] also described a reduced bile acid excretion, with increased dehydroxylation of cholic acid in the gut, possibly favoring bile acid conservation.

In recently completed studies on the steroid excretory pattern of type II hyperlipoproteinemic subjects treated with the soybean diet regimen [Fumagalli et al, 1982], we failed to note any significant change in the pattern of fecal steroid excretion. By monitoring the in- and out-movements of sterols from the plasma and tissues after injection of ^{14}C-cholesterol to the patients, no change in the slope of the plasma specific-activity decay curve could be detected (Fig. 8-VI). The mechanism of the soybean protein diet is thus different from that of drugs affecting steroid excretion into the gut lumen (eg, clofibrate) [Grundy et al, 1972] and of drugs interfering with bile acid reabsorption (eg, cholestyramine) [Nazir et al, 1972]. The lack of marked alterations in the fecal steroid excretory pattern during the first ten days after switching of the dietary protein, in spite of the dramatic plasma cholesterol reduction, is noteworthy.

B. Changes in Amino Acid Composition and Hormonal Secretion

The different amino acid composition of the soybean diet (as a model vegetable-protein diet) and of casein (model animal-protein diet) has been underlined by several authors [Kritchevsky, 1979; Huff et al, 1977]. In particular, Kritchevsky et al [1982] suggested that the high lysine/arginine ratio in casein might inhibit liver arginase, thus reducing the formation of the arginine-rich

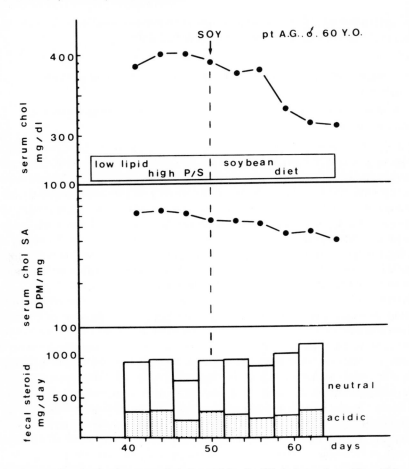

Fig. 8-VI. Changes in plasma total cholesterol levels, C¹⁴-cholesterol specific activity, and fecal steroid excretion in a type II male patient, after switching from a standard low-lipid to the soybean diet. In spite of the marked plasma cholesterol reduction, changes in both the specific-activity decay curve and fecal steroid excretion are negligible [Fumagalli et al, 1982].

proteins (ARP) typical of experimental hyperlipidemic conditions [Rodriguez et al, 1976].

Recent experimental findings [Roberts et al, 1981] have, indeed, shown that ARP is increased when experimental atherogenic diets in rabbits contain casein, whereas it is reduced when soybean is the major dietary protein. Other data by Huff and Carroll [1980], however, do not confirm that the relative arginine richness of the soybean diet may be the major factor reducing choles-

TABLE 8-I. Percent Changes in Plasma Free Amino Acid Levels Before and After Cholsoy Pronto Treatment

	Percent change	P
Arginine	+ 25.8	< 0.01
Lysine	− 5.1	NS
L/A ratio	− 22.8	< 0.01

terolemia. Experiments with different amino acid mixtures in rabbits, eg, 50/50 casein-soy protein added with the essential amino acids from one or from the other diet, did not clearly indicate any dietary amino acid pattern associated with the reduction of cholesterolemia. From these studies, it appears that the predominant protein (soy or casein) is the determinant of the final cholesterolemia. Whatever the case, protein hydrolysates of casein or soybean protein display the same activity on plasma cholesterol as intact proteins [Huff et al, 1977]. In contrast, amino acid mixtures corresponding to the same proteins seem to have a different effect on the rabbit model [Sirtori, 1982].

In type II patients treated with a new preparation of TVP containing a small amount of lecithin (Cholsoy Pronto®) and possibly eliciting a small but significant increase of HDL cholesterol levels [Descovich et al, 1982], the plasma amino acid composition was studied before and after the therapeutic diet. At the end of treatment, the plasma content of free arginine had increased more than 20%, with a corresponding decrease of free lysine. The lysine/arginine ratio was also lowered (Table 8-I). Attempts to correlate these changes in amino acid levels with the reduction of cholesterolemia did not result in statistically significant findings. However, these changes are noteworthy in view of the hypothetical mechanism of the dietary regimen, possibly related to plasma changes in free amino acid or small peptide levels [Sirtori, 1982].

The possibility that the relative arginine richness of the soybean diet might change the pattern of plasma glucagon secretion was evaluated by Noseda and Fragiacomo [1980]. These authors hypothesized that an increase in the relative glucagon (G) and insulin (I) levels, might result in a reduction of cholesterol biosynthesis [Ingerbritsen et al, 1979]. Indeed, in patients chronically treated with the soybean diet, an increased plasma G/I ratio may be observed. Similarly, following a short-term dietary administration, an increased G secretion after arginine stimulus is detected; I secretion, conversely, is not changed. The separation of plasma glucagon into its different molecular weight (MW) forms, further indicated the low MW hormone, exerting the maximal metabolic activity [Srikant et al, 1977], was raised after the soybean diet.

A later experiment compared, in a group of healthy volunteers, the effects of a casein-rich diet, given for one week, with that of the soybean diet, given the following week. The arginine stimulation test was repeated at the end of both weeks. Both diets significantly raised plasma G levels during the stimulation test, compared with the prediet findings. The most likely explanation is that these experimental diets are considerably richer in protein than standard everyday regimens (from approximately 13-14% of calories from protein in the ordinary diet, to 19-20% in the experimental diets). Recent animal studies have suggested that high plasma concentrations of several amino acids induced by high-protein diets might be G secretogogues, rather than arginine per se [Eisenstein et al, 1979]. If the arginine richness of soy protein probably does not provide an adequate explanation, the increase in the low MW form of glucagon in plasma following the diet is of interest, and its significance should be further investigated.

IV. CONCLUSIONS

Administration of TVP, prepared from soybean, to type II hyperlipidemic patients is now an established mode of treatment, capable of markedly reducing elevated plasma cholesterol levels. This effect has been confirmed in both in- and outpatient studies and may be maintained for prolonged periods of time, provided that patients carefully adhere to the regimen. Only minimal plasma cholesterol changes may be expected in subjects with normal cholesterolemia or with borderline plasma cholesterol elevations. The mechanism of this dietary treatment is still obscure. The diet does not induce an enhanced fecal steroid excretion, thus ruling out the possibility that nonabsorbable components may play a significant role. Changes in the plasma free amino acid profile have been detected (markedly reduced lysine/arginine ratio), as well as an increased relative proportion of the low molecular weight form of glucagon in plasma. These last findings need further confirmation and clarification.

ACKNOWLEDGMENTS

This work was supported in part by the Consiglio Nazionale delle Ricerche of Italy (P.F. Medicina Preventiva, Subproject ATS 79 011 13.83). Dr G. Naccari (Gipharmex, Milan) is gratefully acknowledged for the generous provision of Cholsoy.

V. REFERENCES

Descovich GC, Ceredi C, Gaddi A, Benassi MS, Mannino G, Colombo L, Cattin L, Fontana G, Senin U, Mannarino E, Caruzzo C, Bertelli E, Fragiacomo C, Noseda G, Sirtori M, Sirtori CR (1980): Multicenter study of soybean protein diet for outpatient hypercholesterolaemic patients. Lancet 2:709.

Descovich GC, Benassi MS, Cappelli M, Gaddi A, Grossi G, Piazzi S, Sangiorgi Z, Mannino G, Lenzi S (1982): Metabolic effects of lecithinated and non-lecithinated textured soy protein treatment in hypercholesterolemia. In Noseda G, Fragiacomo C, Fumagalli R, Paoletti R (eds): "Lipoproteins and Coronary Atherosclerosis." Elsevier Biomedical Press, p 279.

Eisenstein AB, Strack I, Gallo-Torres H, Georgiadis A, Neal Miller O (1979): Increased gluca gon secretion in protein-fed rats: Lack of relationship to plasma aminoacids. Am J Physiol 236:20.

Fumagalli R, Soleri L, Farina R, Musanti R, Mantero O, Noseda G, Gatti E, Sirtori CR (1982): Fecal cholesterol excretion studies in type II hypercholesterolemic patients treated with the soybean protein diet. Atherosclerosis 43:341.

Grundy SM, Ahrens EH Jr, Salen G, Schreibman PH, Nestel PJ (1972): Mechanism of action of clofibrate on cholesterol metabolism in patients with hyperlipidemia. J Lipid Res 13:531.

Helms P, Gatti E, Sirtori CR (1977): Soybean protein diet and plasma cholesterol. Lancet i:805.

Hodges RE, Krehl WA, Stone DB, Lopez A (1967): Dietary carbohydrates and low cholesterol diets: Effects on serum lipids in man. Am J Clin Nutr 20:198.

Holmes WL, Rubel GB, Hood SS (1980): Comparison of the effects of dietary meat versus dietary soybean protein on plasma lipids of hyperlipidemic individuals. Atherosclerosis 36:379.

Huff MW, Carroll KK (1980): Effects of dietary proteins and amino-acid mixtures on plasma cholesterol levels in rabbits. J Nutr 110:1676.

Huff MW, Hamilton RMG, Carroll KK (1977): Plasma cholesterol levels in rabbits fed low fat, cholesterol-free, semipurified diets: Effects of dietary protein hydrolysates and aminoacid mixtures. Atherosclerosis 28:187.

Huijbregts AWM, Van Schaik A, Van Berge-Henegouwen GP, Van der Werf SDJ (1980): Serum lipids, biliary lipid composition and bile acid metabolism in vegetarians as compared to control subjects. Eur J Clin Invest 10:443.

Ingerbritsen JS, Geelen MJH, Parker RA, Evenson KJ, Gibson DA (1979): Modulation of hydroxymethylglutaryl CoA reductase activity, reductase kinase activity, and cholesterol synthesis in rat hepatocytes in response to insulin and glucagon. J Biol Chem 254:1979.

Keys A, Anderson JT, Grande F (1965): Serum cholesterol response to changes in the diet. I. Iodine value of dietary fat versus 2S-P. Metabolism 14:747.

Kritchevsky D (1979): Vegetable proteins and atherosclerosis. J Am Oil Chem Soc 56:135.

Kritchevsky DL, Tepper SA, Czarnecki SK, Klurfeld DM (1982): Atherogenicity of animal and vegetable protein. Atherosclerosis 41:429.

Liebermeister H, Tolujipur H (1980): Senkung des Cholesterinspiegels durch Zusatz von Pektinen und Sojaprotein zur Reduktionsdiät. Dtsch Med Wochenschr 105:333.

Miettinen TA, Tarpila S (1977): Effect of pectin on serum cholesterol, faecal bile acids and biliary lipids in normolipidemic and hyperlipidemic individuals. Clin Chim Acta 79:471.

Nazir DJ, Horlick L, Kudchodkar BJ, Sodhi HS (1972): Mechanism of action of cholestyramine in the treatment of hypercholesterolaemia. Circulation 46:95.

Nestel PJ, Billington I, Smith B (1981): Low density and high density lipoprotein kinetics and sterol balance in vegetarians. Metabolism 30:941.

Noseda G, Fragiacomo C (1980): Effect of soybean protein diet on serum lipids, plasma glucagon and insulin. In Noseda G, Lewis B, Paoletti R (eds): "Diet and Drugs in Atherosclerosis." New York: Raven Press, p 61.

Potter JM, Nestel PJ (1976): Greater bile acid excretion with soybean than with cow milk in infants. Am J Clin Nutr 32:1645.

Raaij JMA van, Katan MB, Hautvast JGAJ, Hermus RJJ (1981): Effects of casein versus soy-protein diets on serum cholesterol and lipoproteins in young healthy volunteers. Am J Clin Nutr 34:1261.

Roberts DCK, Stalmach ME, Khalil MW, Hutchinson JC, Carroll KK (1981): Effects of dietary protein on composition and turnover of apoproteins in plasma lipoproteins of rabbits. Can J Biochem 59:642.

Rodriguez JL, Ghiselli GC, Torreggiani D, Sirtori CR (1976): Very low density lipoproteins in normal and cholesterol fed rabbits. Atherosclerosis 23:73.

Schwandt P, Richter WO, Weisweiler P (1981): Soybean protein and serum cholesterol. Atherosclerosis 40:371.

Shorey RAL, Bazan B, Lo GS, Steinke FH (1981): Determinants of hypocholesterolemic response to soy and animal protein-based diets. Am J Clin Nutr 34:1769.

Sirtori CR (1982): A "drug" for the prevention of atherosclerosis in vegetable proteins? TIPS 3: 170.

Sirtori CR, Agradi E, Conti F, Gatti E, Mantero O (1977): Soybean-protein diet in the treatment of type II hyperlipoproteinemia. Lancet 1:275.

Sirtori CR, Gatti E, Mantero O, Conti F, Agradi E, Tremoli E, Sirtori M, Fraterrigo L, Tavazzi L, Kritchevsky D (1979): Clinical experience with the soybean protein diet in the treatment of hypercholesterolemia. Amer J Clin Nutr 32:1645.

Sirtori CR, Descovich GC, Noseda G (1981): The soybean protein diet does not lower plasma cholesterol? Atherosclerosis 38:423.

Vessby B, Karlström B, Lithell H, Gustafsson I-B, Werner I (1982): The effects on lipid and carbohydrate metabolism of replacing some animal protein by soy-protein in a lipid-lowering diet for hypercholesterolaemic patients. Human Nutr 36A:179.

Animal and Vegetable Proteins in Lipid
Metabolism and Atherosclerosis, pages 149–168
© 1983 Alan R. Liss, Inc., 150 Fifth Ave., New York, NY 10011

9
Immunological Aspects of Atherosclerosis: The Role of Dietary Protein

Patrick J. Gallagher and Michael J. Gibney

Departments of Pathology and Nutrition, Faculty of Medicine, University of
Southampton, S09 5NH England, UK

I. INTRODUCTION

In the midnineteenth century Rudolf Virchow described the microscopic appearance of atheromatous plaques and emphasized the important role of lipid in the development of lesions [Virchow, 1959]. His views gained ready acceptance and were, in turn, supported by the first series of experimental feeding studies in the early part of the twentieth century [Bailey, 1916; Dock, 1958]. Equally, Virchow can be regarded as the originator of the immunological, or at least the inflammatory, theory of the etiology of atherosclerosis. His descriptions contain clear accounts of the inflammatory infiltrates so often seen in the outer media and adventitia of atheromatous segments of both the aorta and the muscular arteries. These infiltrates are often most prominent in patients dying of atheromatous disease at a relatively young age [Saphir and Gore, 1950] and their density has been related to the severity of the disease [Schwartz and Mitchell, 1962].

Although the inflammatory infiltrates surrounding atheromatous lesions contain macrophages, lymphocytes, and plasma cells, immunoproteins such as complement (C) and immunoglobulins (Ig) are difficult to demonstrate histochemically. However biochemical studies [Hollander et al, 1979] indicate that IgA, IgG, and the third component of complement (C3) are present in higher concentrations in atherosclerotic as opposed to normal arterial intima. Furthermore, studies with radioactively labeled amino acids suggest that immunoglobulin is synthesized in the diseased arterial intima perhaps as part of an immune response to an endogenous or exogenous antigen associated with plaque collagen.

It is now appreciated that several other vascular disorders, such as systemic lupus erythematosus, rheumatoid vasculitis and polyarteritis nodosa have an immunological basis. The essential pathology is deposition of complexes of antigen and antibody in the vessel wall. IgG, IgA, and IgM can be readily demonstrated by immunological techniques but the antigenic component of the complex is often uncertain [MacIver and Gallagher, 1981].

In the same way immune complex vasculitis can be produced experimentally by intravenous injection of foreign protein [Cochrane and Koffler, 1973; Theofilopoulos and Dixon, 1980]. The animal reacts by producing antigen-specific antibodies which in turn combine with the injected protein to form antigen-antibody complexes. Immune complexes are not inherently harmful; but if they lodge in sites such as the renal glomerulus or vessel wall they can activate the complement cascade, and in turn trigger the coagulation system and attract polymorphonuclear inflammatory cells.

A number of different investigators have now demonstrated that animals injected intravenously with foreign proteins and given an atherogenic diet develop substantially more arterial disease than controls fed the diet alone [Minick and Murphy, 1973; Howard et al, 1971; Lamberson and Fritz, 1974]. In particular Sharma and Geer [1977] demonstrated that after an injection of foreign protein both the protein itself and autologous antibody to it could be detected immunohistochemically in the arterial wall. What are taken to be the subsequent series of events are illustrated in Figure 9-I. After immune complexes bind to or are trapped at the endothelial surface, complement is rapidly activated. Components of the complement cascade such as C3a or C5a are powerful chemotaxins, and neutrophil infiltration and platelet aggregation ensue. In these circumstances endothelial necrosis and subsequent thrombus formation are inevitable. Repeated cycles of endothelial injury followed by repair from the underlying media are now considered to be the basis of the development of the early atheromatous lesion [Ross and Glomsett, 1976].

II. THE IMMUNOLOGICAL ROLE OF DIETARY PROTEIN IN ARTERIAL DISEASE

Experiments in both man and animals have shown that a small proportion of dietary protein is absorbed wholly or partially undigested into the mesenteric lymphatics or portal venous system [Alexander et al, 1936; Warshaw and Walker, 1974; Hemmings and Williams, 1978]. The initial contact with a gastrointestinal antigen generates classes of specifically sensitized lymphoid cells in the Peyer's patches of the small intestine. In turn these seed the entire length

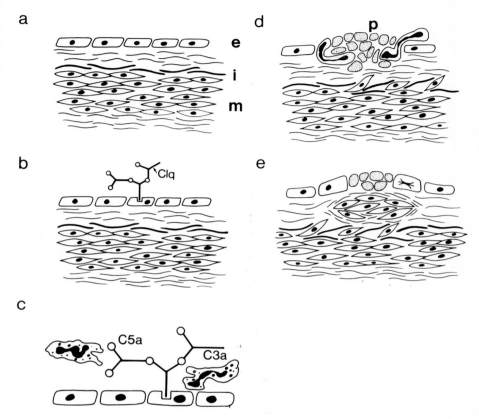

Fig. 9-I. Induction of endothelial injury by immune complexes. a) Normal arterial wall: e, endothelium; i, internal elastic lamella; m, medial smooth muscle cells. b) Immune complexes of antigen (○) and antibody are trapped at the endothelial surface and complement (C1q) is fixed. c) C3a and C5a components of the complement cascade are chemotactic for neutrophil polymorphs (n). d) Endothelial necrosis and platelet thrombus (p) formation ensue. e) Platelet thrombus formation induces smooth muscle proliferation and early intimal thickening [Ross and Glomsett, 1978].

of the gastrointestinal tract and are responsible for the production of antigen-specific IgA [Walker and Isselbacher, 1977]. The antibodies combine with dietary peptides, immobilizing them at the gut surface, preventing their absorption, and increasing their susceptibility to proteolytic degradation [Walker et al, 1977]. Perhaps because of the protection that these secretory immunoglobulins provide, most adults have only very low levels of circulating antibodies to common food proteins [Hunter et al, 1968; Gibney et al, 1980].

Nevertheless, variations in the immunological reaction to dietary antigens could be important in several disease processes.

Patients with coeliac disease have substantially higher levels of antibody to milk protein than healthy controls and the same may be true in severe inflammatory bowel disease [Taylor et al, 1964; Jewell and Truelove, 1972; Jewell and MacLennan, 1973]. Considerable interest was generated by the suggestion that patients with a history of myocardial infarction had increased milk antibody titers [Davies et al, 1974], but subsequent studies failed to confirm this hypothesis [Toivanen et al, 1975; Scott et al, 1976]. In our own study we compared milk antibody levels in patients undergoing angiography prior to coronary bypass surgery with both healthy controls and patients with valvular heart disease. Antibody levels were measured by two different techniques, but there was no evidence that patients with coronary atherosclerosis had higher levels than either of the other groups [Gibney et al, 1980].

If antibodies directed against dietary antigens do play a role in the pathogenesis of atherosclerosis it is more likely to be as components of an antigen-antibody complex rather than as antibody alone (Fig. 9-I). Circulating immune complexes have been described in both clinical and experimental forms of vasculitis [Theofilopoulos and Dixon, 1980; MacIver and Gallagher, 1981] and by two separate groups in patients with ischemic heart disease [Farrell et al, 1977; Füst et al, 1978]. Laboratory methods for the estimation of immune complexes are technically demanding and require rigorous attention to detail. Using both Raji cell immunoassay and a C1q binding test [Jones et al, 1982], we examined sera from patients admitted to a coronary care unit, unselected subjects undergoing coronary angiography, and healthy blood-transfusion volunteers. Irrespective of the final diagnosis, patients from the coronary care and angiography groups had a higher incidence of positive results than non-hospitalized blood-transfusion controls. However, there was no evidence that patients with ischemic heart disease had higher values than those with valvular lesions or that those with genuine acute myocardial infarction differed from the patients with some other cause of chest pain (Table 9-I). We therefore seriously doubt the significance of a single raised level of immune complexes in any hospitalized patient.

In recent experiments we hae investigated the immunological changes following both acute and chronic variations in the pattern of dietary protein intake. Groups of healthy volunteers ate a minimum of four meals per week based on soybean protein, or took a supplement of 700 ml of milk and 300 g of commercial yogurt per day for between 3–6 weeks. Food antigen-specific antibodies were measured by an enzyme-linked immunoabsorbent assay (ELISA) and immune complexes by Raji immunoassay, but no appreciable changes were detected over the experimental period. Volunteers in the milk experiment complained of loss of appetite, and serum cholesterol values fell sig-

TABLE 9-I. Results of Immune Complex Estimations

Group	M:F	Median age (range)	C1q binding Raised	C1q binding Normal	Raji immunoassay Raised	Raji immunoassay Normal
Coronary care unit						
Acute myocardial infarcts	16:6	56 (35–81)	6	16	13	9
Anginal pain	5:2	64 (57–76)	3	4	3	4
Noncardiac pain	5:6	55 (28–67)	2	9	8	3
Coronary arteriography						
Ischemic heart disease	14:5	54 (36–67)	9	10	15	4
Valvular disease	9:12	49 (32–71)	11	10	16	3
Blood-transfusion volunteers	21:29	28 (21–59)	8	42	3	47

nificantly after 3 weeks. Protein intake increased substantially in the soya volunteers and cholesterol levels fell slightly [Goulding et al, 1982].

A second series of experiments concentrated on the acute immunological changes following a high-protein test meal. Dietary antibodies and circulating immune complexes (CIC) were measured at 30-minute intervals after the ingestion of a high energy liquid formulation (5.5 MJ) containing 66 g of milk, soy, or egg protein or a protein-free control. Using the C1q binding test, marked but transient elevations of CIC were detected 30–120 minutes after the milk and egg meals but not with soy or the control (Fig. 9-II). No such changes were seen when complexes were estimated by Raji cell immunoassay. Food antigen-specific antibodies were detected in the plasma of all subjects but showed no consistent pattern of variation in the postprandial period.

Recent experiments in our laboratories [Jones et al, 1982] have shown that whereas C1q binding and Raji immunoassay are efficient at detecting immune complexes with a molecular weight greater than 1×10^6 (> 19S) only C1q binding detects material smaller than this (7–19S). These observations suggest that complexes appearing in the serum of normal individuals after a high-protein meal contain at most only three or four molecules of immunoglobulin. From the results of experiments in mice Haakenstadt and Mannik [1974] concluded that immune complexes of large molecular size were rapidly phagocytosed by the liver and spleen, whereas very small complexes were trapped in the capillaries of the renal glomeruli, the lungs, and the choroid plexus. It is only intermediate-sized particles (7–19S) which can lodge in the walls of arteries, arterioles, capillaries, and venules. Taking these findings together it is at least

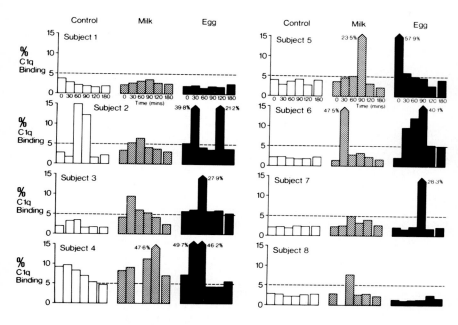

Fig. 9-II. Circulating immune complex levels by C1q binding after a high-protein test meal. The upper limit of normal is 5% binding (horizontal line). Only subject 2 showed a response to the protein-free control meal. Five of eight volunteers reacted at least transiently to 66 g of egg or milk protein.

theoretically possible that the immune complexes which form in normal individuals in the postprandial period could initiate arterial injury.

III. DIETARY ANTIGENS, IMMUNE TOLERANCE, AND EXPERIMENTAL ATHEROSCLEROSIS

Despite the protection afforded by mucosal IgA, a small proportion of dietary proteins, perhaps up to 2% [Warshaw et al, 1977], is absorbed in an immunogenic form. Such continuous challenge provides a persistent, if low-grade, antigenic stimulus to both the gastrointestinal and systemic immune systems. Nevertheless, in the vast majority of adults levels of food antigen-specific antibodies are comparatively low, and as described above quite uninfluenced by acute or chronic variations in dietary protein intake. This suggests that some form of tolerance or hyporesponsiveness has developed to dietary protein, and various mechanisms have been invoked to explain this phenomenon.

There is good clinical and experimental evidence that the liver plays a central role in the phagocytosis of both antigenic material and immune complexes arriving via the portal venous system [Triger and Wright, 1979]. Increased levels of bacterial and dietary antibodies can be detected in patients with advanced liver disease [Triger et al, 1972] and hyporesponsiveness to orally administered antigens is diminished in experimental animals with hepatic cirrhosis [Thomas et al, 1978]. Cell-transfer studies in this animal model suggest that T as opposed to B lymphocytes mediate the hyporesponsiveness to gastrointestinal antigens, and indeed other groups have implicated suppressor T cells in Peyer's patches of the small intestinal mucosa [Asherson et al, 1977]. In contrast André and his colleagues [1975] found that tolerance produced by gastrointestinal challenge could be transferred by a cell-free preparation rich in complexes of antigen and IgA. These findings are of considerable interest in the context of the acute feeding studies described above, suggesting that the transient elevation of antigen-antibody complexes in the postprandial period could play a role in maintaining tolerance to food proteins.

In a series of experimental feeding studies in rabbits we came to suspect that perinatal exposure to a dietary protein was an important event in the induction of tolerance to food antigen. To test this hypothesis we have undertaken a number of dietary studies in different animal species. In the first of these New

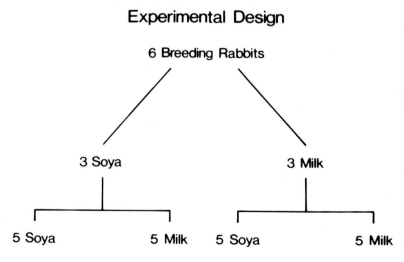

Fig. 9-III.   The design of experiment 1. The fishmeal diets of rabbits were supplemented with either 5% soy or milk protein. After one month the animals were mated. Groups of their weanling offspring (bottom line) were subsequently fed diets based on soy or cows' milk protein (details Table 9-III).

Zealand white breeding rabbits were fed from weaning on a diet in which fish-meal provided the sole protein source. At 3 months of age, as the animals approach sexual maturity, the diet was supplemented with approximately 50 g/kg soy or cows' milk protein. One month later the animals were mated and from their offspring four groups of five animals made up (Fig. 9-III). Two of these groups were given the same dietary protein as their dams, but the others were weaned onto a completely novel diet, ie, soy to milk or milk to soy. Food antibody levels to both proteins were measured by ELISA at weekly intervals (Figs. 9-IV, 9-V). At weaning food antibody levels were low in all animals. After approximately three weeks rabbits given the same protein as their dams formed substantially lower levels of circulating antibodies than those given a different protein. The effect was slightly more marked with soy.

This form of tolerance could be produced in utero by transfer of maternal antibodies across the yolk sac. Alternatively it may be the result of antigenic exposure during that early period of postnatal life when an immunological challenge induces tolerance rather than immunity [Smith, 1961]. Food antigens have been demonstrated in human breast milk [Kulangara, 1980], and we have consistently observed that young rabbits nibble at pelleted diets from 16–18 days of age.

To explore these possibilities a cross-fostering experiment was performed using egg albumin as the dietary protein.

The results (Table 9-II) confirmed the effect found with milk and soy protein: The lowest egg antibodies were in the group bred and suckled from dams fed an egg-containing diet. The cross-fostering part of the experiment showed that some suppression of the antibody response was obtained by both the mammary and uterine routes. The highest levels of antiegg antibodies were in the group reared from and nursed with mothers on an egg-free diet.

Subsequent experiments explored how the magnitude of the immune response to dietary protein influenced the development of experimental atherosclerosis in rabbits.

The diets of the milk and soy-fed rabbits described above were supplemented with 1% cholesterol 30 days after weaning. All animals were killed 90 days later and the macroscopic extent of aortic atherosclerosis estimated by a grid-counting technique. Although the differences in food antigen-specific antibodies persisted throughout the experiment group, mean aortic atherosclerosis scores were similar. Thus although perinatal exposure to a dietary antigen was found to modulate the subsequent immune reaction to food proteins, this phenomenon did not affect aortic atherosclerosis produced by cholesterol feeding.

In experiment 2 atherosclerosis was induced in three groups of rabbits by a cholesterol-free soy protein diet containing 17% saturated fat (Table 9-III). Semisynthetic diets of this type rarely produce serum cholesterol values greater

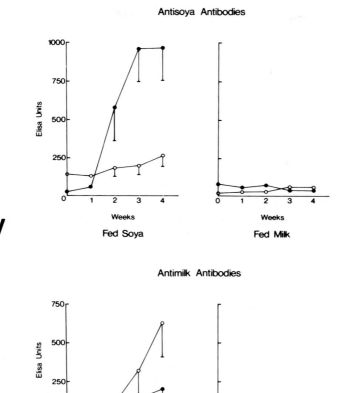

Figs. 9-IV and 9-V.   Food-antibody levels in experiment 1. Open circles: animals reared from dams fed soy. Closed circles: offspring of dams fed milk. Antibody levels were substantially lower in animals fed the same protein as their dams (left-hand graphs). Note uniformly low antibody levels to milk in animals fed soy after weaning and vice versa (right-hand graphs). Antibody levels expressed in arbitrary linear units.

TABLE 9-II. Design and Results of an Experiment to Determine the Relative Importance of the Uterine and Mammary Routes in the Induction of Tolerance to Weanling Rabbits*

	Dams	
Group	Control diet	Egg diet
Diet	(100 g/kg, soy)	(100 g/kg, egg albumin)
Number of animals	3	3
Antiegg IgG[a] (12 weeks)	0.395	2.78
Number of offspring	16	15

	Treatment of weanlings			
Pregnancy	Control dam	Egg dam	Control dam	Egg dam
Lactation	Control dam	Control dam	Egg dam	Egg dam
Diet at weaning	Egg diet	Egg diet	Egg diet	Egg diet
Antiegg IgG at weaning[b]	0.046	0.281	0.028	0.173
1 Month postweaning[b]	1.21	0.76	0.67	0.58

*IgG values in absorbance at 405 nm using the ELISA method of Pathirana et al [1981].
[a]Serum dilution 1:200.
[b]Serum dilution 1:1,000.

TABLE 9-III. Composition of Diets Used in Rabbit Feeding Experiments 1–3 in g/kg

	Soya	Milk	Fishmeal
Soy (Promine D)	332	—	—
Dried skim milk	—	522	—
Casein	—	143	—
White fishmeal	—	—	444
Methionine	6	2	—
Coconut oil	167	148	113
Corn oil	23	37	20
Lactose	278	—	226
Bran	150	150	130
Minerals	69	21	60
Vitamins	9	9	8
Energy (MJ/kg)	18.0	18.0	17.2
Energy from protein	31%	31%	31%
Cholesterol (Soxholet extraction)	0.005%	0.012%	0.012%

than 5 mmol/liter and thus avoid the massive hypercholesterolemia that follows cholesterol supplementation. One set of animals (group 1) was derived from a colony fed a breeding diet containing 5% soy protein continuously for two years. In contrast groups 2 and 3 were bred from a colony fed fishmeal

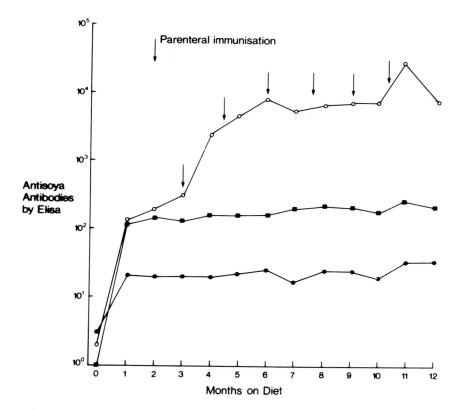

Fig. 9-VI. Group mean antisoy antibodies in experiment 2. Note animals tolerized to soy protein by perinatal exposure (group 1, lowermost line) had substantially less antisoy antibody activity than other groups. Antisoy antibodies expressed in arbitrary units on a logarithmic scale. (● Group 1; ■ group 2; ○ group 3, see text.)

protein. Rabbits in group 3 were immunized parenterally with an emulsified suspension of Freunds complete adjuvant with Promine D at six weekly intervals, whereas in the other groups only saline was used. Group mean live weight gains, energy intake, total cholesterol, and triglycerides were similar in all groups. Animals tolerized to soy protein by perinatal exposure had the lowest levels of circulating antisoy antibodies (Fig. 9-VI) and the least aortic atherosclerosis (Table 9-IV). In addition the tolerized animals had the lowest incidence of raised circulating immune complex levels (Table 9-IV).

The histological appearances of the lesions were broadly similar in all three groups. Atheromatous plaques were most numerous in the ascending aorta

TABLE 9-IV. Results of Rabbit Soy Feeding Experiment 2 (Means ± SEM)

	Group 1	Group 2	Group 3
Maternal diet	Soy	Fishmeal	Fishmeal
Parenteral injection	Saline	Saline	Soy
Total aortic atherosclerosis (%)	2.2 ± 1.5	8.0 ± 2.5	13.2 ± 3.8
Ascending aortic atherosclerosis (%)	10.2 ± 6.2	32.6 ± 9.5	37.3 ± 9.5
Serum cholesterol (mmol/liter)	2.64 ± 0.44	2.61 ± 0.33	2.12 ± 0.23
Serum triglyceride (mmol/liter)	0.96 ± 0.08	0.77 ± 0.06	0.80 ± 0.06
Percent positive immune complex results at 12 months[a]	10%	50%	43%

[a]Tested by C1q binding and Raji cell immunoassay [Jones et al, 1982].

and around the major abdominal branches. The chief features were fibrous intimal thickening, disruption of the medial aortic elastica, dystrophic calcification, and mononuclear and giant cell infiltration (Fig. 9-VII, 9-VIII, 9-IX).

From a pathological standpoint the granulomatous infiltrates seen both in this and in a similar previous study [Gallagher et al, 1978] are both the most interesting and most enigmatic finding. Some of the microscopic features, particularly the aggregation of mononuclear and giant cells around disrupted fragments of elastic tissue are reminiscent of human temporal arteritis. The cause of temporal arteritis is unknown but its rapid and predictable response to corticosteroid treatment suggests an inflammatory basis. Although we have been unable to demonstrate immune complexes immunohistochemically in biopsies from affected patients [Gallagher and Jones, 1982], there is good experimental evidence that intradermal injection of complexes can provoke a granulomatous reaction [Ridley et al, 1982].

Whatever the significance of the histological findings, this experiment does demonstrate that animals made tolerant to a dietary protein are protected from the atherosclerosis induced by cholesterol-free diets containing moderate amounts of saturated fat.

During the course of these experiments data derived from various control groups suggested that fishmeal protein had a substantial and independent effect on the levels of serum cholesterol. To investigate this further, in experiment 3 groups of ten or 15 rabbits were fed diets containing approximately 30% of crude milk, soy, or white fishmeal protein and supplemented with coconut oil (Table 9-III). Mean serum cholesterol and triglyceride values were highest in the fishmeal-fed animals (Fig. 9-X). As was anticipated from previous studies [Carroll, 1978] rabbits given soy had slightly lower values than those fed milk-protein diet. Predictably, the fishmeal animals had substantially more aortic atheroma than those in the other groups (Table 9-V). Histo-

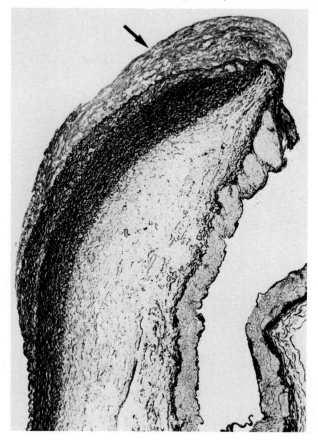

Fig. 9-VII.   Experiment 2. Aorta and coeliac axis. Elastic Van Gieson, × 80. Prominent area of fibrous intimal thickening close to orifice of coeliac axis (arrow). Soy protein diet, maternal diet fishmeal (group 2).

logically fibrous intimal thickening was more pronounced in this experiment than in the previous soy feeding studies. In addition disease was also identified in the carotid and renal arteries (Fig. 9-XI). In a further study (experiment 4) the effect of varying the fat source was studied. Groups of rabbits were given diets containing 445 g/kg of white fishmeal and 133 g/kg of coconut oil (saturated) or maize oil (unsaturated). As in the previous study serum cholesterol values quickly rose to 10 mmol/liter, but values were similar with both diets. After ten weeks of feeding the rabbits were given a stock soy-based laboratory ration, and cholesterol values rapidly reverted to normal or subnormal levels

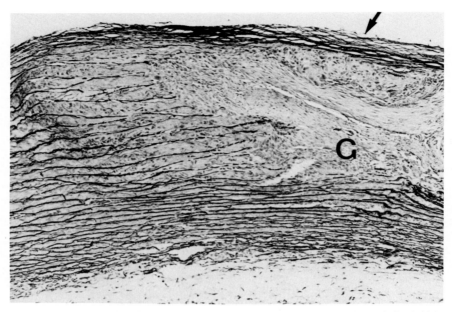

Fig. 9-VIII. Experiment 2. Aorta. Gomori's aldehyde fuschin, × 110. Fibrous intimal thickening (arrow) with disruption of the medial elastic pattern and prominent granulomatous infiltrates (g). Soy-protein diet, maternal diet fishmeal (group 2).

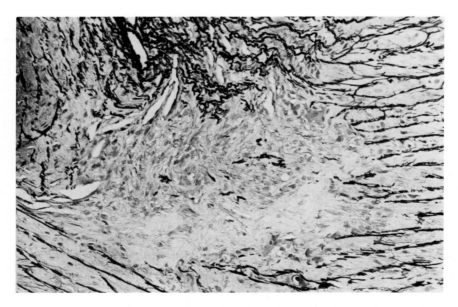

Fig. 9-IX. Experiment 2. Aorta. Gomori's aldehyde fuschin, × 320. The lamella pattern is disrupted and there is a giant cell and mononuclear response to fragments of elastica. Soy-protein diet, parenteral soy immunization, maternal diet fishmeal (group 3).

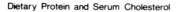

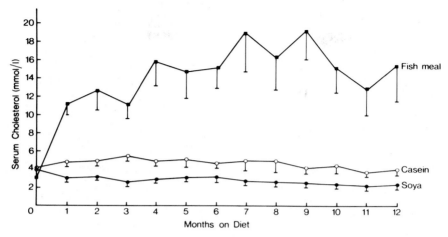

Fig. 9-X.   Total serum cholesterol levels in experiment 3 (means ± SEM).

TABLE 9-V. Percent Aortic Involvement by Atherosclerosis in Experiment 3 (Means ± SEM)

	Soy	Milk	Fishmeal
Ascending aorta[a]	29.8 ± 6.7	38.6 ± 6.4	77.0 ± 6.8
Total aorta	8.9 ± 2.4	14.6 ± 4.3	75.4 ± 7.9

[a]Aortic valve to ductus arteriosus scar.

(Fig. 9-XII). These results indicate that the hypercholesterolemia induced by diets rich in fat and white fishmeal is independent of the degree of saturation of the oil.

There is evidence from previous feeding studies [Kritchevsky, 1979] that the dietary ratio of lysine to arginine might influence serum cholesterol and atherosclerosis. Indeed, recent data from Kritchevsky et al [1982] have shown a linear relationship between the severity of rabbit aortic and thoracic atherosclerosis and the lysine:arginine ratios of diets based on fish protein, casein, or cows' milk protein. Our studies do not corroborate this hypothesis. Rabbits fed fishmeal had markedly more aortic atheroma than rabbits fed soy, as would be predicted from their relative ratios of lysine to arginine (0.95 vs 0.85). However, rabbits fed cows' milk protein had considerably less atheroma than fishmeal-fed animals, in spite of milk having a higher lysine:arginine ratio than fishmeal (1.22 vs 0.95). Clearly these data suggest that the lysine: arginine hypothesis requires further study to ensure that comparisons between protein sources are not confounded by pre- or postnatal tolerization.

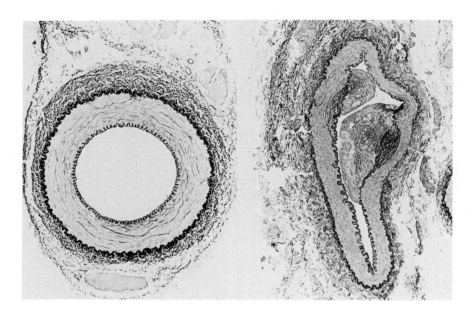

Fig. 9-XI. Experiment 3. Intimal thickening in the renal artery of a fishmeal-fed rabbit (right-hand side) contrasted with normal appearance (left). Elastic Van Gieson, × 65.

As an alternative hypothesis to explain the hypercholesterolemic effect of dietary protein, the potential role of insulin and glucagon should be considered. These hormones are directly linked to the activity of rat hepatic 3-hydroxy-3-methyl-glutaryl coenzyme A (HMG-CoA), especially to diurnal variations in its activity [Lakshaman et al, 1973]. Furthermore, since insulin is a potent regulator of extrahepatic lipoprotein lipase, it can make a significant contribution to the turnover of very low density lipoprotein. We investigated the effects of dietary protein source on the postprandial secretion of insulin in healthy male volunteers who consumed test meals (4.2 MJ) containing (1) isolated soy protein, glucose, and butter; (2) casein, glucose, and butter; and (3)

Fig. 9-XII. Mean total serum cholesterol values in experiment 4. Animals were fed the fishmeal diets for ten weeks (between arrow heads) and then transferred to a stock laboratory diet.

Fig. 9-XIII. Serum insulin (μ/ml) following the consumption of a test meal (4.2 MJ) by five healthy male volunteers: (●) casein; (○) isolated soybean protein, glucose, and butter; (△) glucose and butter.

XII

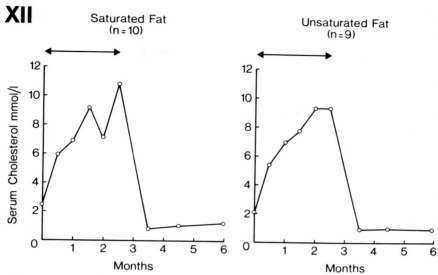

XIII

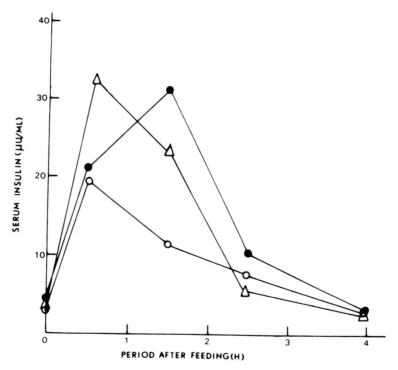

glucose and butter. The energy distribution between protein, fat, and carbohydrate in the protein-containing meals was 1:2:1. The results are shown in Figure 9-XIII. Insulin secretion in the soy-fed volunteers matched that of the protein-free controls while casein produced a significant reduction in the mean level of insulin in serum ($P < 0.05$). Amino acids, particularly arginine, are known to act as powerful secretogogues of insulin [Fajans et al, 1967; Eisenstein and Strack, 1978]. Thus the release of arginine following the feeding of arginine-rich soy may have provided an additional stimulus to insulin secretion. Clearly, much further work is required to investigate the likelihood of this hypothesis of insulin/glucagon involvement.

IV. CONCLUSIONS

We conclude from our studies that dietary protein can make a significant contribution to atherosclerosis in rabbits. However, this capacity is dependent upon the ability to develop a systemic antibody response to the dietary protein. Our experiments failed to demonstrate increases in serum systemic antibodies in healthy volunteers fed soy- or milk-rich diets, in either the short or long term. This confirms previous studies in which levels of antimilk antibodies in patients with coronary artery disease were no greater than in controls free of arterial disease. However, there was substantial evidence from our experiments that in healthy volunteers there is a marked rise in circulating immune complexes to potentially pathological levels, following animal-protein-rich meals. Taken together these data indicate that the role of dietary protein in the immunological induction of arterial diseases is potentially great.

ACKNOWLEDGMENTS

We acknowledge the collaboration of Cecily R. Casey, N.J. Goulding, Karen Jones, D.B. Jones, Chitra Pathirana, G.P. Sharratt, and T.G. Taylor in various of the experiments described above. Financial support was provided by the Wessex Regional Health Authority and Nestle Nutrition S.A. The manuscript was prepared by Miss Margaret Harris and Mrs Jennifer Goulding.

V. REFERENCES

Alexander HL, Shirley K, Allen D (1936): The route of ingested egg white to the systemic circulation. J Clin Invest 15:163–167.
André C, Heremans JF, Vaerman JP, Cambiaso CL (1975): A mechanism for the induction of immunological tolerance by antigen feeding: Antigen antibody complexes. J Exp Med 142:1509–1519.

Asherson GL, Lembala M, Perera MACC, Mayhew B, Thomas VR (1977): Production of immunity and unresponsiveness in the mouse by feeding contact sensitizing agents and the role of suppressor cells in the Peyer's patches mesenteric lymph nodes and other lymphoid tissues. Cell Immunol 33:145-155.

Bailey CH (1916): Atheroma and other lesions produced in rabbits by cholesterol feeding. J Exp Med 23:69-85.

Carroll KK (1978): Dietary protein in relation to plasma cholesterol levels and atherosclerosis. Nutr Rev 36:1-5.

Cochrane CG, Koffler D)1973): Immune complex disease in experimental animals and man. Adv Immunol 16:185-264.

Davies DF, Rees BWG, Johnson AP, Elwood PC, Abernethy M (1974): Food antibodies and myocardial infarction. Lancet 1:1012-1014.

Dock W (1958): Editorial: Research in arteriosclerosis — The first fifty years. Ann Intern Med 49:699-705.

Eisenstein AB, Strack J (1978): Amino acid stimulation of glucagon secretion by perfused islets of high-protein fed rats. Diabetes 27:370-376.

Fajans SS, Floyd JC, Knopf RF, Conn JW (1967): Effect of amino acids and proteins on insulin secretion in man. Rec Prog Horm Res 23:617-662.

Farrell C, Bloch B, Nielsen H, Daugharty H, Lundman T, Svenag J-E (1977): A survey for circulating immune complexes in patients with acute myocardial infarction. Scand J Immunol 6:1233-1240.

Füst G, Szondy E, Szekely J, Nanai I, Gero S (1978): Studies on the occurrence of circulating immune complexes in vascular diseases. Atherosclerosis 29:181-190.

Gallagher PJ, Jones K (1982): Immunohistochemical findings in temporal arteritis. Arthritis Rheum 25:75-79.

Gallagher PJ, Muir CA, Taylor TG (1978): Immunological aspects of arterial disease. Atherosclerosis 30:361-363.

Gibney MJ, Gallagher PJ, Sharratt GP, Benning HS, Taylor TG, Pitts JM (1980): Antibodies to heated milk protein in coronary heart disease. Atherosclerosis 37:151-155.

Goulding NJ, Gibney MJ, Gallagher PJ, Morgan JB, Jones DB, Taylor TG (1982): The immunological consequences of a high intake of soya-bean protein in man. Qualitas Plantarum (in press).

Haakenstad AO, Mannik M (1974): Saturation of the reticuloendothelial system with soluble immune complexes. J Immunol 112:1939-1948.

Hemmings WA, Williams EW (1978): Transport of large breakdown products of dietary protein through the gut wall. Gut 19:715-723.

Hollander W, Colombo MA, Kirkpatrick B, Paddock J (1979): Soluble proteins in the human atherosclerotic plaque. Atherosclerosis 34:391-405.

Howard AN, Patelski J, Bowyer DE, Gresham GA (1971): Atherosclerosis induced in hypercholesterolaemic baboons by immunological injury; and the effects of intravenous polyunsaturated phosphatidyl choline. Atherosclerosis 14:17-29.

Hunter A, Feinstein A, Coombs RRA (1968): Immunoglobulin class of antibodies to cows' milk casein in infant sera and evidence of low molecular weight antibodies. Immunology 15:381-388.

Jewell DP, MacLennan ICM (1973): Circulating immune complexes in inflammatory bowel disease. Clin Exp Immunol 14:219-226.

Jewell DP, Truelove SC (1972): Circulating antibodies to cows' milk protein in ulcerative colitis. Gut 13:796-801.

Jones DB, Goulding NJ, Casey CR, Gallagher PJ (1982): C1q binding and Raji immune complex assays: A comparison using defined immunoglobulin aggregates. J Immunol Methods 53:201-208.

Kritchevsky D (1979): Vegetable protein and atherosclerosis. J Am Oil Chem Soc 56:135–140.

Kritchevsky D, Tepper SA, Czarnecki SK, Klurfeld DM (1982): Atherogenicity of animal and vegetable protein: Influence of the lysine to arginine ratio. Atherosclerosis 41:429–431.

Kulangara AC (1980): The demonstration of ingested wheat antigens in human breast milk. IRCS Med Sci 8:19.

Lakshaman MR, Nepokroeff CM, Ness GC, Dugan RE, Porter JW (1973): Stimulation by insulin of rat liver hydroxy-methylglutaryl co-enzyme A reductase and cholesterol synthesizing activities. Biochem Biophys Res Commun 50:704–710.

Lamberson HV, Fritz KE (1974): Immunological enhancement of atherogenesis in rabbits: Persistant susceptibility to atherogenic diet following experimentally induced serum sickness. Arch Pathol 98:9–16.

MacIver AG, Gallagher PJ (1981): The pathology of arterial disease. Br J Anaesth 53:675–687.

Minick CR, Murphy GE (1973): Experimental induction of atheroarteriosclerosis by the synergy of allergic injury to arteries and lipid rich diet. II. Effect of repeatedly injected foreign protein in rabbits fed a lipid rich cholesterol-poor diet. Am J Pathol 73:265–300.

Pathirana C, Gibney MJ, Taylor TG (1981): The effect of dietary protein source and saponins on serum lipids and the excretion of bile acids and neutral sterols in rabbits. Br J Nutr 46:421–432.

Ridley MJ, Marianayagam Y, Spector WG (1982): Experimental granulomas induced by myobacterial immune complexes in rats. J Pathol 136:59–72.

Ross R, Glomsett JA (1976): The pathogenesis of atherosclerosis. N Engl J Med 295:369–377, 420–425.

Saphir O, Gore I (1950): Evidence for an inflammatory basis of coronary arteriosclerosis in the young. Arch Pathol 49:418–426.

Schwarz CJ, Mitchell JRA (1962): Cellular infiltration of the human arterial adventitia associated with atheromatous plaques. Circulation 26:73–78.

Scott BB, McGuffin P, Swinborne ML, Losowsky MS (1976): Dietary antibodies and myocardial infarction. Lancet 2:125–126.

Sharma HM, Geer JC (1977): Experimental aortic lesions of acute serum sickness in rabbits. Am J Pathol 88:255–266.

Smith RT (1961): Immunological tolerance of non-living antigens. Adv Immunol 1:67–129.

Taylor KB, Truelove SC, Wright R (1964): Serologic reactions to gluten and cows' milk proteins in gastrointestinal disease. Gastroenterology 46:99–108.

Theofilopoulos AN, Dixon FJ (1980): Immune complexes in human diseases: A review. Am J Pathol 100:529.

Thomas HC, Ryan CJ, Benjamin IS, Blumgart LH, MacSween RNM (1978): The immune response in cirrhotic rats: The induction of tolerance to orally administered protein antigens. Gastroenterology 71:114–117.

Toivanen A, Viljanen MK, Savilahti E (1975): IgM and IgG anti-milk antibodies measured by radioimmunoassay in myocardial infarction. Lancet 2:205–207.

Triger DR, Alp MH, Wright R (1972): Bacterial and dietary antibodies in liver disease. Lancet 1:60–63.

Triger DR, Wright R (1979): Immunological aspects of liver disease. In Wright R, Alberti KGMM, Karran S, Millward-Sadler GH (eds): "Liver and Biliary Disease." London: WB Saunders, pp 182–196.

Virchow R (1859): "Cellular Pathology. As Based upon Physiological and Pathological Histology," 2nd Ed. New York: Dover Publications Inc, 1971. (translated by Chance F, 1860).

Walker WA, Isselbacher KJ (1977): Intestinal antibodies. N Engl J Med 297:767–773.

Walker WA, Wu M, Bloch KJ (1977): Stimulation by immune complexes of mucus release from goblet cells of the rat small intestine. Science 197:370–372.

Warshaw AL, Bellini CA, Walker WA (1977): The intestinal mucosal barrier to intact antigenic protein. Am J Surg 133:55–58.

Warshaw AL, Walker WA (1974): Intestinal absorption of intake (sic) antigenic protein. Surgery 76:495–499.

Index

A

Age
  and cholesterolemic response to
      casein diets, 30–32
  and hypercholesterolemia in
      humans, 128
Albumin, egg. *See* Egg albumin
Alfalfa, 11
Amino Acid(s)
  composition of diet
    changes in, and effects on plasma
        cholesterol, 143–146
    and cholesterol turnover, 63–69
    effect on serum apolipoproteins,
        74–78
    and human hypercholesterolemia,
        129
    and plasma cholesterol levels,
        12–14
    and serum cholesterol, 25–26
    see also Cholesterol, plasma,
        effects of amino acid
        composition
  effects on cholesterol metabolism,
      9–15
  L-, 12-14
  supplementation, and cholesterol
      metabolism in swine, 102–103
Animal protein
  atherogenic component, 2
  and coronary heart disease, 20
  effect on cholesterol absorption, 55
  see also specific proteins

Antibody
  food, 155–160
  milk protein, 152
Antigen
  dietary, and immune tolerance,
      154–166; *see also* Immunoglobulins
  plaque collagen, 150
Apolipoprotein(s), serum
  apo A_I and B, 74–78, 118–119
  apo E, 70, 72, 74
  effect of casein in humans, 120
  effect of dietary protein and amino
      acid mixtures, 74–78
  and hypocholesterolemic efficacy of
      soy protein, 75–79
  LCAT activity, 75
  SDS-gel electrophoresis, 39
  in VLDL and LDL, 74, 76–78
  VLDL apo E, 38
  see also Lipoprotein(s)
Arginine
  and casein in hypercholesterolemia,
      23–24
  effects on triglyceridemia, 90
  /lysine ratio, *See* Lysine/arginine
      ratio
  and plasma cholesterol, 69, 70
  and plasma glucagon, 145–146
  -rich proteins, 144
  and VLDL and LDL, 67
Assay
  Clq binding, 152, 153, 154
  ELISA, 152, 156
  immuno-, Raji cell, 152, 153